W. H. Davenport Adams

Everyday Objects

Picturesque Aspects of Natural History

W. H. Davenport Adams

Everyday Objects
Picturesque Aspects of Natural History

ISBN/EAN: 9783337025618

Printed in Europe, USA, Canada, Australia, Japan

Cover: Foto ©Lupo / pixelio.de

More available books at **www.hansebooks.com**

EVERYDAY OBJECTS.

(*Frontispiece.*)

EVERYDAY OBJECTS

OR

Picturesque Aspects of Natural History.

WITH NUMEROUS ILLUSTRATIONS.

EDITED AND ENLARGED BY

W. H. DAVENPORT ADAMS,

AUTHOR OF "THE CIRCLE OF THE YEAR," "SWORD AND PEN," "BEFORE THE CONQUEST," ETC.

> "To know
> That which about us lies in daily life,
> Is the prime wisdom."
> MILTON.

WILLIAM P. NIMMO:
LONDON: 14 KING WILLIAM STREET, STRAND;
AND EDINBURGH.
1876.

PREFACE.

THE very favourable reception accorded both by Press and Public to the "Circle of the Year," has induced me to prepare a second volume, similar in design, but dealing with different branches of the same subject. As the former was founded on the *first* series of a popular French work, "Les Saisons," by M. Hoefer, so the present has been suggested by the *second* series; but in availing myself of it, I have omitted much, I have revised more, and at various parts my additions have been considerable. And here, as in my former effort, I have written from a popular rather than a scientific point of view. It has not been my object to sketch the outlines or lay down the foundations of any science; but to show, as best I could, how much of wonder and beauty enters into our daily life, and what inexhaustible sources of study lie at our very feet. It is, perhaps, a misfortune of our common systems of education that they too much neglect the tuition of the eye; that the young are not taught to mark the curious and interesting objects which are comprehended within their daily vision; that they know so much about ancient mythology and so little about modern science,—so much about gods and heroes, so little about stars and flowers.

I have called this volume "Everyday Objects," not because those which it describes may be seen every day, but because they mostly belong to the region of the commonplace and familiar; and I have called it "Picturesque Aspects of Natural History," because I have endeavoured, in companionship with my French *collaborateur*, to indicate the poetical side of the various sciences into which I have presumed to penetrate. If it should awaken a love of nature in any breast, or develop a spirit of inquiry, which may lead the student further and further on the path of knowledge, the labour bestowed upon these pages will not have been in vain.

The instinct of curiosity,—says M. Hoefer, in his preface to the first series of "Les Saisons,"—is the awakening of the intellectual life: it commences with the lisping of the child, accompanies the adult in every phase of his existence, and, far from becoming extinct with the last throb of the heart, revives before the unknown shadows of the grave. What, then, is there in the whole world of greater importance to follow and direct than the movements and impulses of this curiosity, of these uncertain pulsations of the soul? In this lies the secret of all education; and upon education depends the future of humanity.

Unfortunately, he continues, the methods hitherto employed have been absolutely insufficient. And the insufficiency is most notable as regards the imperfect and defective training given to the instinct of curiosity. Observe the child. Of everything which excites his attention, he never fails to ask you the *reason why*. It is thus that he enters into the connexion of "cause" and "effect." It is a sign. But instead of following up this natural indication, and developing the thought by the exercise of the reason, we proceed as if the being under our charge were incapable of reason; we overload the memory of the child with a multitude of words, whose value he cannot understand until later in life, and perhaps never. The true direction of the mind is to proceed from the thought to the word, and not from the word to the thought. It is for want of having recognised and applied this principle that our educational systems have failed so utterly.

Let us take, for example, the study of nature. No science, assuredly, ought to prove more attractive to the mind than natural history. Yet mark how repulsive zoology, botany, and mineralogy are made at the very outset, by the dryness of their nomenclatures and the dreariness of their classifications. Undoubtedly, it is necessary to lay down a course of study in the midst of the marvels which everywhere surround us; undoubtedly names are required for the objects which attract our notice. But are not the methods we employ directly opposed to the end we set before ourselves?

I address myself to parents and teachers; and I say to them, Do you wish to inculcate a love of science, and yet put into the hands of your children or pupils books which differ as widely from the book of nature as human brotherhood—(a fiction!)—differs from universal gravitation? Instead of familiarising us at first with the animals and plants within our everyday reach, you collect, under the same irrevocable iron "form," genera and species never intended to meet in any one particular zone, and many of which are so rare that few persons will ever be fortunate enough to see them except in collections and engravings. And, curious to state, the rarest species nearly always obtain your preferences; judging, at least, from the minute descriptions which you consecrate to them. Monstrous absurdity! You seek at a distance that which lies close to your hands, as if the Everyday Objects above, beneath, and around, were unworthy of the science you profess.

But here we must pause. Upon the principles thus laid down by M. Hoefer, have been founded the two unpretending companion volumes, of which the second is now submitted to the lenient judgment of the public.

W. H. DAVENPORT ADAMS.

CONTENTS.

BOOK I.—WINTER.

CHAP.   PAGE

I. WHAT MAY BE SEEN IN THE HEAVENS :—

The Number of the Stars,	4
The Great Bear and the Little Bear,	8
Orion,	13
Diurnal Movement,	15
Determination of the Cardinal Points,	17

II. WHAT MAY BE SEEN UPON THE EARTH :—

The Snow,	32
Red Snow,	39
The Eternal Snow,	44
The Inhabitants of the Eternal Snows,	48
The *Arvicola Leucurus*,	49
The Marmot,	53
The Chamois,	56
The Eagle and the Wren,	57
The Snow Bunting,	66
The Red-billed Crow,	68
Reptiles,	70
Inferior Animals,	71

WHAT MAY BE SEEN ON THE EARTH, *continued* :—

	PAGE
Herbaceous Plants which best endure the Cold of Winter,	75
The Dog Mercury,	77
The Garden Nightshade,	82
The Dog's-tooth Grass,	88

BOOK II.—SPRING.

CHAP.
I. WHAT MAY BE SEEN IN THE HEAVENS :—

The Earth's Figure is seen in the Sky as in a Mirror,	102

II. WHAT MAY BE SEEN ON THE EARTH :—

Causes of the Circulation of the Sap,	132
The Daisy,	138
The Tulip,	152
The Heliotrope,	156
The Anemones,	157
The Arum,	161
The Ranunculaceæ,	165
The Wood-louse,	169
The Dragon-flies,	174

BOOK III.—SUMMER.

I. WHAT MAY BE SEEN IN THE HEAVENS :—

The Adumbrated Sphere,	191

II. WHAT MAY BE SEEN ON THE EARTH :—

The Perianth,	208
The Calyx,	208
The Corolla,	223
The Prunella,	230
The Scutellaria,	235
The Lilies,	241

	PAGE

WHAT MAY BE SEEN ON THE EARTH, *continued :*—

The Gentians,	250
An Alpine Excursion,	256
The Pimpernel,	260
The Mole—The Staphylinus—The Mole Cricket,	265
The Earwig,	278

BOOK IV.—AUTUMN.

CHAP.

I. WHAT MAY BE SEEN IN THE HEAVENS :—

The Circle, and the uniform Movement of the Stars (according to the Theory of the Ancients destroyed by Kepler),	289
The Solar Constitution,	292
Result of recent Astronomical Researches,	296

II. WHAT MAY BE SEEN ON THE EARTH :—

Chemical action of Light,	312
Action of Heat,	313
Arable Land,	318
Mushrooms or Agarics,	325
The Number of Vegetable Species distributed over the whole Surface of the Globe,	337
The Harvest Bug,	349
The Cheese Mite,	354
The Number of Animal Species distributed over the whole Surface of the Globe,	356
What is Chlorophyll?	366
Carnations and Pinks,	371
The Eglantine and the Convolvulus,	379
Metamorphosis : a Physico-philosophical Meditation,	384

APPENDIX, ... 405

BOOK I.

WINTER.

Lastly, came Winter clothèd all in frieze,
Chattering his teeth for cold that did him chill;
Whilst on his hoary beard his breath did freeze,
And the dull drops, that from his purpled bill
As from a limbeck did adown distil:
In his right hand a tippèd staff he held,
With which his feeble steps he stayèd still;
For he was faint with cold, and weak with eld;
That scarce his loosèd limbs he able was to wield.
—SPENSER, *The Faerie Queene*, Canto vi.
(Of Mutability).

You naked trees, whose shady leaves are lost,
Wherein the birds were wont to build their bower,
And now are clothed with moss and hoary frost,
Instead of blossoms, wherewith your buds did flower;
I see your tears that from your boughs do rain,
Whose drops in dreary icicles remain.
—SPENSER, *The Shepherd's Calendar*,
Eclogue i.

CHAPTER I.

WHAT MAY BE SEEN IN THE HEAVENS.

> Skies flower'd with stars,
> Violet, rose, or pearl-hued, or soft blue,
> Golden, or green, the light now blended, now
> Alternate.
> —P. J. BAILEY, *Festus.*

OUR observation of the celestial phenomena may most easily be made in the winter-time. Then the nights are long, and the vault of heaven is crowded with stars, and, unilluminated by the moon, exhibits all its splendours. In the other seasons of the year, and particularly in summer, the twilight gleam encroaches, so to speak, upon a portion of the nights, which are otherwise so brief, and precludes our vision from any exact estimate of the stars. Those demi-tints, those soft subdued reflections of light, scarcely permit the eye to distinguish even stars of the first and second magnitude, which shine like spots of dull gold on a background of pale silver.

The Number of the Stars.

How many are the stars?

To such a question comes the immediate answer, They are infinite in number.

But, after a little meditation, we begin to perceive that the question, apparently so simple, is, in reality, one of very great complexity. Let us endeavour to disentangle its various threads.

We must not forget that, in every scientific analysis, it is important we should, in the first place, separate two intimately united elements,—the individual who observes, and the product of the observation. The former, the "sensorial factor," is subject to every condition of space and time; the second, the "intellectual factor," tends, by its generalisations, to free itself from those very conditions which are the inseparable co-efficients of matter and movement. The individual passes; save from an outer standpoint, we know not whence he comes, nor whither he goes. The product of the observation remains; transmissible from generation to generation, it will gradually expand and increase, if it be founded upon truth; but, on the contrary, its splendour will wane, and will eventually disappear, if it be founded upon error. Eternal is this spectacle of actors and puppets succeeding one another uninterruptedly upon the same stage! As one falls, another steps forward into his place, and so the great army marches forward with unbroken ranks.

He who, "in cities pent," sees the sky only through a

garret window, or in the narrow intervals between house and house, can form no accurate idea of the magnificence of the firmament. The peasant, the shepherd, or the labourer, spent with his daily work, prefers sleep to astronomical vigils; and even amongst those more favoured sons of fortune, who enjoy sufficient leisure, but few are found who feel a genuine pleasure in the study of the stars. Though they are the poetry of heaven, their music is inaudible to the majority of souls. We content ourselves with an occasional careless glance at their serene loveliness, and then turn again to the pleasures or avocations of commonplace life.

But, come; let us arouse ourselves! Let us quit the city for awhile; let us throw off all thought of its too-engrossing pursuits; let us find time to count the stars. Gentle readers, I ask you to follow me.

Ah, me! how small is the train of followers! How great my delusion in supposing that a complete phalanx of students of the celestial wonders would reply to my invitation!

We have now arrived in the open country; and here, on the summit of this gentle ascent, crowned with a clump of leafless trees, we pause. The sky glitters with a cold, keen light, which is reflected back by the snowy plains. While the eye ranges delightedly over the starry vault, the ear is struck by the distant sound of bells, which, at the midnight hour, ring in the infant year—ring in so many hopes and expected joys, and unexpected sorrows—ring out so many passing pleasures and rudely dissipated visions.

As the chime glides softly over the meadows, and along the

resounding vales, and through the leafless woods, repeated by echo after echo, until its music dies away in the distance,

FIG. 1.

like our recollections of the dreams of youth, we murmur to ourselves that solemn song of the poet, which so aptly blends the regrets of the past with the anticipations of the future; we exclaim—

> "Ring out, wild bells, to the wild sky,
>   The flying cloud, the frosty light :
>   The year is dying in the night ;
> Ring out, wild bells, and let him die.
>
> "Ring out false pride in place and blood,
>   The civic slander and the spite ;
>   Ring in the love of truth and right,
> Ring in the common love of good.
>
> "Ring out old shapes of foul disease ;
>   Ring out the harrowing lust of gold ;
>   Ring out the thousand wars of old,
> Ring in the thousand years of peace.
>
> "Ring in the valiant man and free,
>   The larger heart, the kindlier hand ;
>   Ring out the darkness of the land,
> Ring in the Christ that is to be." *

The spectacle is majestic and impressive. Let us seek, in the first place, to ascertain our position in reference to the four points of the compass—the four cardinal points. But how is this to be done? By day it is easy enough. I have only to turn myself towards the sun when it has reached the highest point of its diurnal course, and there, in front of me, lies the south, in my rear the north, the east on my left, and on my right the west.

But is it possible to ascertain one's position during the absence of the "orb of day?"

* Tennyson, "In Memoriam," cvi.

both possible and easy, provided the sky be clear and cloudless.

But this condition is as necessary by day as by night. How can we determine in which direction lies the south, if the sun be hidden from our gaze by an uniformly opaque atmosphere, and if objects, lit up by a diffuse light, project no shadow at any time of the day?

Endeavour to group together the stars which more particularly strike your gaze; and be careful, in these groupings, to define every fantastic figure which is suggested by your vivid imagination. Undoubtedly, our earliest ancestors, the "world's gray forefathers," proceeded in this manner, in their anxiety to lay hold of some definite guiding-marks in yonder ocean of sparkling atoms. And to study a science by its history is to follow up its successive development.

The Great and the Little Bear.

Observe yonder very remarkable group of seven stars; nearly all are of the same splendour, and they are so arranged as to figure an antique chariot, provided with a somewhat curved axle pole.

Observe it carefully. And not far from this group you will detect another, by no means so conspicuous, but exactly resembling it in form. This second chariot is turned in an inverse direction, and the stars composing it, with three exceptions, are much less brilliant.

Here, then, are two groups of stars, clearly distinguished by their configuration—two *constellations*, for such is the scientific name given to all the stellar groups.

It has been the fortune of the first of these two groups to strike the eye of the most indifferent observer from the remotest antiquity; and its likeness to a *quadriga* early procured it the name of a *car* or *chariot*. For those Christians who pleased themselves in studding the sky with Biblical personages, it is *David's Chariot*. This species of apotheosis was borrowed from the Pagans. They placed in the skies their divinities, their demigods, their heroes, and the principal facts and stories of their mythology. For the Greeks and Romans the "Chariot of David" was the female of the Bear, an *ursa*, or ἀρκτός. Whence came this transfiguration? Listen to the fanciful old myth.

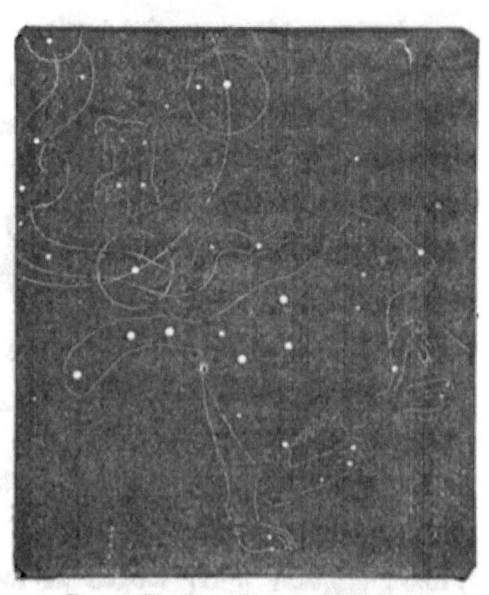

Fig. 2.—The Great Bear and Little Bear.

Callisto was the most beautiful of the daughters of the King Lycaon. Jupiter, who may appropriately be styled the "Don Juan" or "Lovelace" of the heathen Olympus, fell in love with her; and she bore him a son, named Arcas, who gave his name to Arcadia, that land of song and fable, groves and streams, where Lycaon exercised his sovereign sway.

Juno, the queen of heaven, and wife of the so-called king of gods and men, transported by her jealous rage, changed Callisto into a she-bear; who, one day, would have been unwittingly slain by Arcas, if Jupiter, opportunely appearing on the scene, had not metamorphosed the hunter into another animal, *Ursa Minor*, or the Little Bear. According to this myth, the Little Bear will be but a transformation of the former, who was the Great Bear, or, before all and above all, *the* Bear.

It is somewhat surprising, according to certain writers, that Homer should refer to only one of these constellations :—

"Ἀρκτοῖσ᾽ ἥν καὶ ἄμαξαν ἐπίκλησιν καλέουσιν.*
(The Bear, which men the Chariot also name).

But the learned commentators who have censured the poet for making no distinction between Ursa Major and Ursa Minor, probably never looked at the starry vault with an attentive eye; otherwise, like all the world, they might have convinced themselves that the seven stars, *septem triones* (whence the word "septentrion"), forming the beautiful constellation, which, undoubtedly, long before Homer's time, was known as "The Bear" or "The Celestial Chariot," were all that could be seen. With a single exception, these stars are of the second magnitude—that is to say, they, so far as regards their brilliancy, rank next to the most brilliant stars of the firmament. The least conspicuous star in the group—

* Homer, "Odyssey," v. 273.

one of the third magnitude—occupies the base of the pole of the Celestial Chariot, or of the Bear's tail; it is the fourth star counting from the extremity of the tail. On celestial charts, it is particularised by the fourth letter of the Greek alphabet, δ (delta).

Observe, in passing, that the first of these charts, wherein the stars of a constellation were indicated by Greek characters, appeared in 1603, at Augsburg, under the title of "Uranometria." Its author, Jean Bayer, an amateur astronomer, who died in 1660, conceived the idea of designating by the first letters of the Greek alphabet—α, β, γ, δ, and so on—the most noticeable stars. The animals bearing the names of the constellations are drawn in this map with very considerable care; but it requires, let us hasten to add, much imagination and good-will to recognise, in the form of a stellar group, the animal shown in the drawing.

Thus far Ursa Major. The four stars of the quadriga, or chariot, have been employed to form the dorso-lumbar region of the animal; the three others define its tail; and, finally, twenty-four little stars, some of which are hardly visible to the naked eye, compose the head and paws of the celestial "plantigrade."

As for Ursa Minor, it is impossible to distinguish it immediately when you are unaccustomed to surveying or examining the celestial vault. To detect its position, you require to be forewarned of it; to know, in the first place, that there exists in the vicinity of the Bear an exactly similar stellar group. The point of the tail—α in Ursa Minor—alone possesses a splendour comparable to that of the principal stars in Ursa Major. But how construct a figure with one star? The four other stars, two of which mark the anterior part of the animal's body, and two others the tail, properly so called, are only of the third magnitude: they are marked β, γ, δ, ι. Finally, the stars which define the posterior portion, marked ζ and η on Bayer's chart, are only of the fourth magnitude—in other words, are scarcely visible. The eye, to detect them, must be wholly free from any gleam of light.

Many generations passed before they succeeded in discovering what a single individual solved during his brief career. All Homer's contemporaries, and, prior to these, tens of millions of mortals, had contemplated the sky, and yet none of them had detected the difference between Ursa Major and Ursa Minor. The distinction, therefore, is of a comparatively recent date; probably does not date back earlier than the sixth century before the Christian era.

Let us recall ourselves, now, to the question propounded. The first impression produced by the aspect of the sky during a beautiful winter night is, we repeat, that the number of the stars is infinite. This wholly spontaneous thought, which, to some extent, imposes itself on the mind long before the reason attempts any calculation, is, strange to say, both false and true.

But how can a thought be both false and true? Nothing is easier than to explain the seeming contradiction. We shall return to it hereafter, after we have indulged in some indispensable digressions.

Orion.

One of the finest and loftiest flights of Longfellow's imagination is to be found in his poem on the occultation of Orion. He has seldom, if ever, sounded a more vigorous strain. After alluding to that music of the spheres which Pythagoras dreamed of, and which Shakespeare has described in a passage of great beauty, he continues :—

> "Beneath the sky's triumphal arch
> This music sounded like a march,
> And with its chorus seemed to be
> Preluding some great tragedy.
> Sirius was rising in the east;
> And, slow ascending one by one,
> The kindling constellations shone.
> Begirt with many a blazing star,
> Stood the great giant Algebar,
> Orion, hunter of the beast!
> His sword hung gleaming by his side;
> And, on his arm, the lion's hide
> Scattered across the midnight air
> The golden radiance of its hair."

The most ancient observer who wished, with his own eyes, to assure himself whether the number of the stars was infinite, must have quickly perceived that, in spite of an apparent impossibility, it is no difficult task to complete their enumeration. To execute this operation conveniently, however, we must invent a process; and of all processes, the simplest, and that which first occurs to the mind, is to group the stars by configurations which, to a certain degree, are *mnemo-technical*.

Such, in our belief, is the true origin—a point so often and laboriously discussed—of the *asterisms* or *constellations*. Their fanciful, mythological, or poetical embellishments, are of later date.

The census or enumeration of the stars, which we suppose to have commenced during our winter nights, must at first have been limited to the most characteristic groups, composed of the most brilliant points. In this scientific labour the first rank would necessarily be occupied by Arctos (or Ursa Major) and Orion. Why? Because these two constellations attract and rivet everybody's gaze.

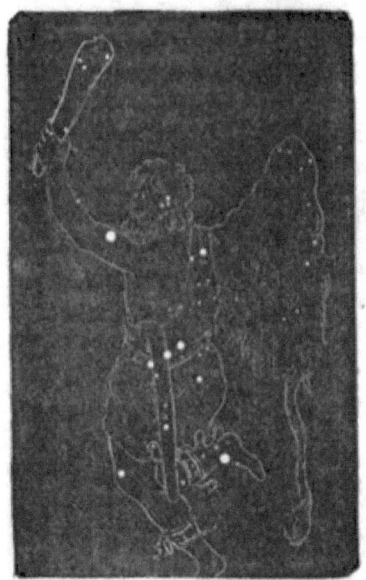

Fig. 3.—Orion.

Orion is situated on the side opposite to the Great Bear. It is the most beautiful constellation in our western sky. You may easily recognise it by three stars, very close together, which are inscribed, as it were, in the centre of a great trapezium of four stars, two of which are of the first magnitude. Beneath the three first stars, called the Three Kings, or Orion's Belt, is visible a small stellar group of the fourth and fifth magnitude, near which, with a good average glass, may be distinguished the largest and most remarkable of the nebulæ.

Here we find the mythologists—those theologians of the Greco-Roman polytheism—at disagreement. According to an ancient legend, immortalised by Homer—

> "Aurora sought Orion's love, . . .
> Till, in Ortygia, Dian's wingèd dart
> Had pierced the hapless hunter to the heart." *

The giant, in the lower world, is still animated by a burning passion for the chase—

> "There huge Orion, of portentous size,
> Swift through the gloom a giant-hunter flies ;
> A ponderous mace of brass, with direful sway,
> Aloft he whirls, to crush the savage prey ;
> Stern beasts in trains that by his truncheon fell,
> Now, grisly forms, shoot o'er the lawns of hell." †

According to later traditions, the giant Orion, son of Tura and Neptune, was endowed by his father with the faculty of walking upon the sea as well as upon earth. He abandoned himself to the fierce joys of the chase in the wooded isle of Crete, to whose shades he had accompanied Diana and Latona. Swollen with pride, he defied to combat all the monsters of the universe, and was slain by a scorpion which the earth had engendered under his feet. But, through the intercession of Diana, a place was given to him in the firmament opposite Scorpio.

DIURNAL MOVEMENT.

Let us put aside these dreams of the world's youth, and return to the reality.

---

* Homer, "Odyssey," Book v., 121-124.
† Ibid., Book xi., 571-574.

Nature, transformed by the ancients into a multiple divinity, never fails to overwhelm with surprise the observer who interrogates her with simplicity and without any preconcerted system. And it was thus that he who first undertook to enumerate the stars, by the help of the constellations, made at once the greatest and most unexpected discovery. What, in fact, was not his astonishment on seeing the gradual displacement of objects which, at the first glance, appeared immovable!

To this very natural astonishment soon succeeded, we doubt not, a desire to analyse the phenomenon. The most beautiful constellations of the firmament, Ursa and Orion, will have their points of repery on the star-gemmed sphere. An attentive study, eagerly pursued through a certain lapse of time, would teach him that Orion rises and sets like the sun and the moon, while the Bear, remaining perpetually above the horizon, neither rises nor sets. Stimulated by curiosity, the observer would afterwards assure himself that the whole of the celestial vault revolved upon an axis, while the stars divided into groups; remain fixed, fixed in this sense, that they constantly maintain among themselves the same relations of distance. The idea of a solid sphere, to which the stars were attached like golden nails, then came quite naturally to the human mind. Such, undoubtedly, was the origin of the discovery of *diurnal movement;* of that general movement which carries all the stars from west to east, to bring them back to the same points in the course of one complete day.

To hear our professors of astronomy invariably repeating,

that "the spectator of the starry vault may see, every moment, new stars rising above the horizon,—may see them mount the sky,—halt in their upward march when they have attained a certain elevation,—afterwards re-descend, and pass below the horizon;"—to hear, we say, these words incessantly reproduced, one would think that a cursory glance at the sky would suffice to reveal the general movement, and that what is within the ken of the first comer, should not be called a discovery.

But we see in this another of those illusions which blind contemporaries as to the time-long efforts of their predecessors to discover the very results which long ago became our common patrimony. Unquestionably, if you have eyes, you cannot fail to see the apparent movement of the earth and moon; but from thence to the relation of the whole celestial sphere is a wide interval. How many men are there who possess, on the one hand, sufficient patience to fix their gaze only for a couple of hours on the same point of the starry firmament; and, on the other, sufficient intelligence to estimate the relation of this point to a fixed point of the horizon, and to measure, by the thought, the interval separating these two points? Let each one ask himself.

Determination of the Cardinal Points.

However it may be, the discovery of the rotation of the celestial system must have been rapidly brought to perfection as it was transmitted from one generation to another. It must soon have been recognised that this sphere is inclined in such a manner that one of its *poles*—the *poles of the world*, which, in reality, are simply the prolonged extremities of the axis of terrestrial rotation—is always above the horizon, while the other remains below. And this phenomenon would lead to the geometrical conception of an axis of rotation of the celestial sphere. Thus we may explain, with perfect ease, why the Bear and the neighbouring constellations should describe perfect circles, and the other and more distant constellations only arcs of circles, of a greater or lesser diameter; finally, without even looking at the sky, we can understand that some stars there are which show themselves on the horizon, only to disappear immediately, and others which remain completely invisible to the inhabitants of our climates. By a singularly fortunate coincidence, the pole, that geometrical point around which revolve those circumpolar constellations that are continually above our horizon, is occupied by a star "well known to fame," and hence, on the faith of its renown, supposed by many people to be a star of peculiar brilliancy.* It is named the *Polar Star* (α in Ursa Minor), and is between the second and third magnitude.

Now if, with arms extended, we so place ourselves that our back shall be turned to Polaris, we shall have opposite to us the point of the arc occupied by the sun at *noon;* on our left the east, and on our right the west. It is thus we may easily learn our position in the absence of the orb of day.

The discovery of this simple mode of guidance was, nevertheless, an epoch in history. From thence the mariner grew

---

* The pole-star, however, is not the true polar point, but distant from it about 1° 32', which, in A.D. 2100, will be decreased to 26' 30".

bold enough to quit the coast, which he had hitherto hugged with timorous prudence, and venture out into the open sea. Thenceforth, the darkness disappeared; new countries were revealed to one another, and nations, which from time immemorial had remained apart, were brought into frequent communication.

It was with eyes fixed upon the Bear, which alone does not bathe itself in the waters of Ocean, that Ulysses set out from Calypso's enchanted island.

According to Homer, who reflects in his immortal work the condition of scientific knowledge among his contemporaries, the ocean was a great broad river, surrounding the earth with circumfluent volume, and in its waves the stars were bathed or extinguished in the evening, to be rekindled in the morning on the opposite side.

By saying that the Bear alone does not bathe in the waters of Ocean *—

$$\text{Οἴη δ' ἄμμορός ἐστι λοετρῶν Ὠκεανοῖο}—$$

the poet plainly shows that Ursa Minor, and the other circumpolar constellations, were unknown in his time.

If the knowledge of these constellations was from the beginning so useful and so necessary to navigation, the constellation nearest to the pole could not, at first, have served as a guide to any but a people essentially maritime. And here we find the Phœnicians, or Tyrians, in the foremost rank.

After reminding us that Ursa Major was also called Helice,

---

* Homer, "Odyssey," Book v., 275.

or "the spiral," as in the famous passage in the "Argonauta" of Apollonius Rhodius,—

> "Night in the east poured darkness; on the sea
> The wakeful sailor to Orion's star
> And Helicè turned heedful,"—

and Ursa Minor, Cynosura,—that is, the dog's tail,—Manilius,[*] a Latin poet, who wrote at the beginning of the Christian era, goes on to say:—

"At one of the extremities of the world's axis are two constellations, well known to the hapless mariner: they are his guides when the bait of gain impels him across the ocean. Helice is the larger, and describes the larger circle: it is recognised by its seven stars, which rival one another in splendour; and by this it is that the Greeks steer their barks. The smaller, Cynosura, describes a lesser circle: it is inferior both in size and lustre; but, according to the witness of the Tyrians, is of greater importance. For the Phœnicians no safer guide exists when they seek to approach a coast invisible from the high seas."

The testimony of Manilius is confirmed by that of Aratus and Strabo. The pseudo-Eratosthenes, in his book on the constellations, refers to Ursa Minor under the name of Φοινίκη, the "Phœnician." It appears, then, to be established that the Phœnicians were the first to group a constellation of the same general outline as Helice, the Little Bear, or Ursa Minor. But that, as we have already explained, the two constellations do not lie in the same direction, every one may see:

[*] Manilius, "Astronomicon," Book i., 291 · 300.

> "Nec paribus positæ sunt frontibus; utraque caudam
> Vergit in alterius rostrum, sequiturque sequentem."*
>
> Not in the same direction do they face :
> The one its tail towards the other's snout
> Turns, and they thus, pursuing, each pursue.

Certain it is that the Phœnicians, as experienced seamen, would guide their course by the constellation lying nearest to the pole. But was this constellation the same which we now-a-days call Ursa Minor? It is quite allowable for us to put such a question, because everybody knows that, owing to the movement of the terrestrial axis around the poles of the ecliptic, the axis of the world (the terrestrial axis prolonged) is displaced to an extent which becomes perfectly appreciable at the end of a certain time.† We may calculate, therefore, that the pole, now situated, as we have already said, near the star Polaris (α in Ursa Minor), was formerly at some distance from it. So, at the epoch of the greatest prosperity of the Phœnician people, or about three thousand years ago, the north pole would nearly correspond with a star in Draco, now 24° 52′ distant.

[This constellation is shown in fig. 1, between Ursa Major and Ursa Minor; the α in Draco is a star surrounded by a circle, like the Polar Star, α in Ursa Minor.]

That the constellation of Draco was well known to the

---

* Manilius, "Astronomicon," i. 308, 309.

† This displacement amounts to about fifty seconds annually, or, more accurately, to 50″.3. It is easy, therefore, to calculate that the complete rotation of the terrestrial axis around the poles of the ecliptic will occupy 25,765 years.

ancients, we may gather from a passage in the "Phenomena" of Aratus, a work partly translated by Cicero:—

"The Dragon, like the sinuous course of a river, uncoils his long scaly body, and surrounds with undulating folds the two constellations of Ursa Major and Ursa Minor."

Bringing together these different facts for the sake of comparison, we arrive at the conclusion that the Polar Star, by whose scintillating light the early mariners steered their tiny keels, was not the Polaris of to-day—α in Ursa Minor—but α in the constellation of the Dragon.

The Arabs, those navigators of the Waterless Sea (as they poetically designate the desert of Sahara), have bestowed particular appellations on several stars; but they guide themselves rather by their radiance than by their position. Thus, such stars as α Draco, α Cepheus, α Cygnus, which have occupied, and, in the course of centuries, will again occupy the place of Polaris, have received no special denomination; while the stars of Ursa Major, α and β (occupying the posterior angles of the chariot), are called Dubke and Merak;* γ, δ, ε, ζ, η, which follow in due succession—Phegæa, or Phad, Megrez, Alioth, Mizar, and Ackaïr, or Benetnasch. Certain stars in the same constellation, which are barely visible, have also received distinctive names: such is Alcor, a star between the fifth and sixth magnitude, in the tail of Ursa Major, between Mizar

---

* These are also called the Pointers, because an imaginary line from the lower to the upper, prolonged in the same direction, passes nearly over the Polar Star.

and Benetnasch. This star, it is true, had a special use: it served the Arabs as the test of a good eyesight.

A further proof that the Arabs founded their stellar nomenclature almost exclusively upon the lustre and colour of the stars, is obvious in the names which they gave to the stars forming the constellation of Orion. (See Fig. 2.) Thus, α and β, two stars of the first magnitude, occupying the right or eastern shoulder, and the left or western foot of the giant-hunter, are called respectively, Betelguese and Rigel; the star γ, named Bellatrix, in the left shoulder, is of the second magnitude, like the stars δ, ϵ, ζ, which represent Orion's Belt, and bear the names of "the Three Kings" and "St James's Staff." Now the star η, marking the right knee or inferior eastern angle of the brilliant trapezium, is only of the third magnitude; therefore, it has received no special designation.

The colour by which some stars are distinguished could not have failed to be remarked by those observers who first began to enumerate, or take census of, the celestial bodies. Thus Sirius, the most refulgent of the stars of heaven, situated in Canis Major, is of a bluish-white, like Rigel; and Arcturus, situated on the prolongation of the tail of Ursa Major, is reddish-yellow, like Betelguese.

Sirius, or the Dog-star, rose heliacally at the hottest time of the year, and hence the Greeks were accustomed to ascribe all the diseases of the season to its influence. It was—

"The star
Autumnal; of all stars, in dead of night,
Conspicuous most, and named Orion's dog:

> Brightest it shines, but ominous, and dire
> Disease portends to miserable man."

To sum up: the figurative grouping of the stars, the variety of their luminous magnificence, their position towards Polaris, the determination of that position by the longitudinal circles passing through the axis of the world, and twisted perpendicularly to this axis by the circles parallel to the Equator, — such is the aggregate of the elements which must, at a very early period, have presided over the enumeration of those sparkling points, each of which is the centre of a system.

Finally, are there any stars which the eye cannot perceive? Such a question would never have been propounded to the ancients. And why? Because no reasoning would have drawn from them an admission that it was possible by artificial means to enlarge the range of our eyesight. They would have deemed it madness to pretend to improve and develope what is not of human creation; the visual apparatus, as it is bestowed on us by nature, they supposed to be the most perfect instrument which man could imagine. And, in truth, nothing could fairly be objected to this way of looking at things.

The 48 constellations (21 northern, 12 zodiacal, and 15 austral) indicated by Ptolemœus, contain a total of 1026 stars, whose relative positions had been determined by Hipparchus. To undertake an enumeration of the stars, and to transmit the result to posterity, appeared to Pliny an audacity before which even a god would have recoiled (*Hipparchus—ausus, rem etiam Deo improbam, annumerare posteris stellas*).*

Yet numerous doubts had already risen in the mind of Hipparchus as to the accuracy of the number recognised. In the first place, the ancients undoubtedly knew, as we do, that the visual faculty is not the same in all individuals; that there are some who, in the same celestial space, see more stars than others. Many persons can discern up to stars of the seventh magnitude, while with others the sight fails far within that limit. The ancients must also have known, as we do, that, for the enumeration to be complete, the sky must be observed from all the points of the terrestrial surface on which man is planted. Even in our own days the catalogues of the southern heavens are far from being perfect. Finally, more than two thousand years before the time of Galileo, Democritus had already enunciated the opinion that the Milky Way was a mass of innumerable stars. All these signs should have been accepted as warnings against premature affirmations.

The invention of telescopes suddenly enlarged the question, and it became necessary to establish a line of demarcation between the number of stars visible to the naked eye and the number visible through the agency of the telescope. Argelander, the author of the "Uranometria," has found that the stars visible to the naked eye, over the entire surface of the heavens, range from 5000 to 5800. Otto Struve, employing Herschel's method of computation, has estimated at upwards

* Pliny, "Historia Naturalis," Book ii., 24.

of twenty millions (20,374,034) the number of stars visible with the Herschel 20-feet telescope.

But, in presence of all the nebulæ resolvable into stellar masses, and before the development of the artificial range of our sight,—in presence, finally, of that hopeless perspective which the more we discover the more we perceive how

FIG. 4.

much there remains to discover,—are we not forcibly carried back to our point of departure?

Ought not the imagination which, at the first glance, led us to believe the number of stars to be infinite,—ought it not to draw us nearer to the truth?

How should the imagination reveal to us, without difficulty, what the intellect, assisted by the senses, can only discover after ages of assiduous exertion?

These questions, it seems to us, are worthy of our studious consideration.

We subjoin a table of the constellations in both hemispheres, with the number of stars in each, for the convenience of our younger readers.

NORTHERN CONSTELLATIONS.

Ursa Minor, the Lesser Bear,	24
Ursa Major, the Great Bear,	87
Perseus, and Head of Medusa,	59
Auriga, the Charioteer,	66
Boötes, the Herdsman,	54
Draco, the Dragon,	80
Cepheus,	35
Canes Venatici, the Greyhounds Asteria and Chara,	28
Cor Caroli (Heart of Charles II.),	3
Triangulum, the Triangle,	16
Triangulum Minus, the Lesser Triangle,	10
Musca, the Fly,	6
Lynx,	44
Leo Minor, the Lesser Lion,	53
Coma Berenicis, Hair of Queen Berenice,	43
Cameleopardalis, the Giraffe,	58
Mons Menelaus, Mt. Menelaus,	4
Corona Borealis, the Northern Crown,	21
Serpens, the Serpent,	64
Scutum Sobieski, Sobieski's Shield,	8
Hercules, with the Dog Cerberus,	113
Serpentarius, or Ophiuchus, the Serpent-Bearer,	74
Taurus Poniatowski, Poniatowski's Bull,	7
Lyra, the Harp,	22
Vulpeculus et Anser, the Fox and Goose,	37
Sagitta, the Arrow,	18
Aquila, the Eagle, with Antinous,	71
Delphinus, the Dolphin,	18
Cygnus, the Swan,	81
Cassiopeia, the Lady in her Chair,	55
Equulus, the Horse's Head,	10
Lacerta, the Lizard,	16
Pegasus, the Flying Horse,	89
Andromeda,	11
Tarandus, the Rein-deer,	12

SOUTHERN CONSTELLATIONS.

Phœnix,	13
Apparatus Sculptoris, the Sculptor's Tools,	12
Eridanus Fluvius, the River Po,	84
Hydrus, the Water-Snake,	10
Cetus, the Whale,	97
Fornax Chemica, the Chemical Furnace,	19
Horologium, the Clock,	12
Rheticulus Rhomboidialus,	10

Xiphias Dorado, the Sword-Fish, ...	7
Celapraxitellis, the Engraver's Tools, ...	16
Lepus, the Hare, . . .	19
Columba Noachi, Noah's Dove,	10
Orion, ...	78
Argo Navis, the Ship Argo, .	64
Canis Major, the Great Dog, .	31
Equulus Pictoris, . . .	8
Monoceros, the Unicorn, .	31
Canis Minor, the Lesser Dog,	14
Chameleon, ...	10
Pyxis Nautica, the Mariner's Compass, ...	4
Piscis Volans, the Flying-Fish, ...	8
Hydra, the Serpent, . .	60
Sextans, the Sextant, . .	41
Robur Carolinum (Charles II.'s Oak), ...	12
Antlia Pneumatica, the Air-Pump, ...	3
Crater, the Cup, . . .	31
Corvus, the Crow, . . .	9
Crux, the Cross, . . .	6
Apis Musca, the Bee or Fly, .	4
Avis Indica, the Bird of Paradise, ...	11
Circinus, the Mariner's Compass, ...	4
Centaurus, the Centaur, . .	35
Lupus, the Wolf, . . .	24
Norma, or Euclid's Square, .	12
Triangulum Australe, the Southern Triangle, . . .	5
Ara, the Altar, . . .	9
Telescopium, the Telescope, .	9
Corona Australis, the Southern Crown, ...	12
Pavo, the Peacock, . .	14
Indus, the Indian, . . .	12
Microscopium, the Microscope,	10
Octans Hadliensis, Hadley's Octant, ...	43
Grus, the Crane, . . .	14
Toucan, the American Goose,	9
Piscis Australis, the Southern Fish, ...	24
Mons Mensa, the Table Mountain, ...	30

ZODIACAL CONSTELLATIONS.

Aries, the Ram, . . .	66
Taurus, the Bull, . . .	141
Gemini, the Twins, . . .	85
Cancer, the Crab, . . .	83
Leo, the Lion, . . .	95
Virgo, the Virgin, . . .	110
Libra, the Balance, . .	51
Scorpio, the Scorpion, . .	44
Sagittarius, the Archer, . .	69
Capricornus, the Goat, . .	51
Aquarius, the Water-Bearer, .	108
Pisces, the Fishes, . . .	113

CHAPTER II.

WHAT MAY BE SEEN UPON THE EARTH.

> Ah, bitter chill it was!
> The owl, for all his feathers, was a-cold;
> The hare limped trembling through the frozen grass,
> And silent was the flock in woolly fold.
>
> —KEATS.

HE winter of 1867–68 will count among the severest recorded in meteorological annals. As early as the winter solstice the cold began to make itself felt. In a few days the centigrade thermometer sank to 12° below zero, through the influence of a very keen north-east wind. At Paris, where, on an average, the winter temperature is two degrees higher than in the surrounding country, the Seine was completely frozen for upwards of a fortnight. To meet with a similar phenomenon we must go back as far as 1788. In January 1830, when, on the 17th, the temperature sank down to 17°.3, the Seine was also frozen, but the ice speedily melted. The extreme cold of

1788 coincides, like that of 1830, with the appearance of two comets. In bringing together these and other similar facts, some writers are induced to believe themselves authorised in establishing theories which attribute a certain frigorific influence to the comets. But no such coincidence existed in the winter of 1867-68, nor in any other years signalised by the occurrence of excessive frost.

What are we to think of the supposed influence of the moon upon the weather?

This question, so constantly revived, is here not out of place. The exceptionally prolonged cold, during which the thermometer remained for three weeks below zero, the barometer oscillating between 76° and 76°.5, commenced on the 22d of December, three days before the new moon; now, it is on Christmas-day, at 48 min. past 11 p.m., that the moon is found in conjunction,—that is to say, has become completely invisible to us by passing between the earth and the sun. And the thaw, which terminated this period of frost, commenced on the 12th of the following January, just three days after the full moon; the exact moment of its opposition, when the moon reflected upon us the whole hemisphere of its borrowed lustre, took place on the 9th, at 2 min. past 11 in the evening. It is then in the neighbourhood of the syzygies (conjunction and opposition) of the moon that we must place the commencement and termination of the cold period to which we have been alluding.

We should not have thought of recalling these coincidences,

if it had not occurred to us that some meteorologists, in accordance with the popular belief, have attributed to the syzygies a marked influence on the changes of the weather. Toaldo has deduced from half-a-century's observations, taken at Padua, this general fact, that the maximum of influence manifests itself at the syzygies, and somewhat more at the new than at the full moon; that the minimum coincides with the first and second quarter; that the action of the perigee (*minima* distance of the moon from the earth) is equal to that of the full moon; and that the action of the apogee (*maxima* distance of the moon from the earth) is double that of the quarters. Observe that the Italian meteorologist extended this influence to three days before and three days after a phase, for the moon's passage through the syzygies; while he restricted it to a day before and a day after, for the quadratures.

The work which Toaldo did for the climate of Padua, Pilgram had already executed for that of Vienna. But the result at which he arrived, after five-and-twenty years of observation (from 1763 to 1788), was the contrary to that of Toaldo: namely, that the new moon is the least active of all the phases in reference to changes of weather. What, then, are we to conclude? That the problem is one of extreme difficulty, and that there are probably several elements necessary to its solution, which at present escape us. Then, too, we ought to have a clear understanding of what is meant by "changes of weather;" we must eliminate all vagueness from the word, and not allow it to be governed by any preconceived theory.

The Snow.

> Thick clouds ascend, in whose capacious womb
> A vapoury deluge lies, to snow congealed.
> Heavy they roll their fleecy world along,
> And the sky saddens with the gathered storm.
> Through the hushed air the whitening shower descends;
> At first thin wavering; till at last the flakes
> Fall broad, and wide, and fast, dimming the day
> With a continual flow.
> —THOMSON, *The Seasons.*

The earth is covered with snow; it is enveloped, as the poets say, in a shroud of white. But this phrase, poetical as it may appear, is, in reality, inadmissible. A shroud is used to wrap round a dead body, a corpse, whose elements, since they are no longer maintained united by the undefinable principle of life, go to form other compounds,—more permanent and lasting,—which will mingle with the earth, the water, and the air. But the earth which the snow covers preserves, on the contrary, the germ of life in the seeds and roots of plants; it rests itself, only for the purpose of communicating, at the return of spring, a new impulse to the sap, whose circulation sleeps during winter.

The moment is propitious for studying the snow: come, then, let us examine it.

And, first, what is snow? Put a little into the hollow of your hand, and see what transpires.

It melts, and leaves nothing but water as a residuum.

Snow, then, is frozen water,—water which existed in the atmosphere in the state of vapour, and which, to speak the language of physicists, has passed from the gaseous state into the liquid, and thence into the solid. If you doubt its identity with water, let a chemist analyse a portion of it for you: he will tell you that it is composed, like distilled water, of hydrogen and oxygen, in the pro-

FIG. 5.—A Snowy Landscape.

portion of two parts of the former to one part of the latter. The reader will, of course, understand that we abstract all foreign substances which may accidentally have got mixed up with it.

It was once a wide-spread opinion that snow is favourable to vegetation, on account of the salts which it contains. Analysis, however, gave a negative result; it demonstrated the absence of these salts. Recourse was then had to another hypothesis: it was supposed that the air contained in snow is richer in oxygen than the free air, and that to the action of this gas must be attributed its fertilising property. Another error! The truth really is, that snow maintains the soil which it covers at a perceptibly constant temperature, and that, when thawing, it mellows it by its aqueous infiltrations; so that if, before a fall of snow, the earth has experienced the action of a strong frost capable of killing injurious insects, all the chances will be in favour of a fertile year.

Snow forms crystals. To observe them clearly, you must examine the snow which falls in very cold and dry weather. It then appears to be a dust composed of little thin plates. Look at the small flake which has fallen on your coat-sleeve; it is isolated; hasten to examine it before it melts, or before other flakes become amalgamated with it. What a graceful star! (Fig. 6, *a*). It is formed of six regular rays. There are others which have only three, four, or five rays. But on inspecting these more closely, you see that many of these rays are broken or abortive, and that, when finally analysed, each star possesses the same number of rays.

Why are there continually six rays? Why are there never more nor fewer than this number? One might suspect in nature a peculiar affection for the number *six:* as, for example, in the cells of the bees and the wasps, which form a regular hexagon (Fig. 6, *b*). Why, in the infinity of polygons, has the instinct of these insects only chosen one hexagon? What is the reason for this preference?

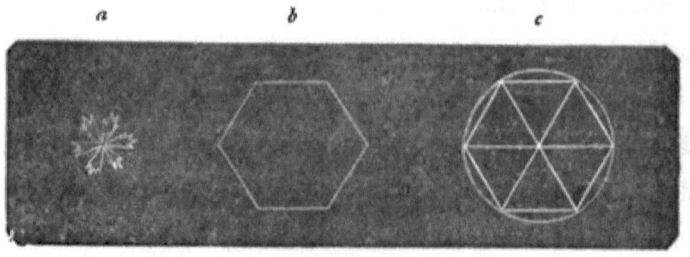

Fig. 6.

If you interrogate geometry, it will reply to you that, of all the polygons inscribed in a circle (Fig. 6, *c*), there is but one whose sides are equal to the radius of that circle; and this polygon is exactly that of the bee and wasp's cell. Here, then, is a very singular coincidence. If you afterwards examine very minutely the work of the bee, you will find in each cell of the honeycomb a pyramidal base, composed of three equal rhombs, whose angles solve a grand geometrical problem, that of giving the maximum of space with the minimum of matter. The papier-maché combs of the wasp are formed of a single row of cells, each of which has a nearly level bottom. This is all that is required; for these cells are destined, not for the reception of honey, but only of the larvæ, the offspring of their architects.

Do not think that you have but to pick up a thumbful of snow to procure your crystals! These change their form very quickly, and it is almost impossible to detect it in snow which has remained for any length of time upon the ground. The great flakes which fall in relatively mild weather, when the temperature borders upon freezing-point, are often nothing better than masses of small amorphous atoms of ice; to get at the crystals, you must remove the kind of icy varnish which encases them.

For the accurate observation of the crystallisation of water which precipitates itself in the air, we have at our disposal a means as simple as convenient—a pane of glass. All we have to do is to arrange everything in such a manner that the congelation shall be both slow and certain; on this condition alone can we obtain well-defined crystals. A cold room is best adapted for this kind of experimentation; and thus you will frequently see deposited upon the window-glass, in an uninhabited chamber, some exceedingly graceful designs, as follow.

Fig. 7.

These are asteriæ, — arborescent, and leaf-like crystals,— imitating the beautiful foliage of ferns and mosses. The severer the cold, the more regular, be it understood, is the formation of these crystals.

Owing to its dazzling whiteness, snow is a great reflector of light, and singularly illuminates the darkness of the winter nights. The long dreary nights of the polar world are lit up by the glories of the magnetic auroras, joined to the radiancy of the snow. This induces us to repeat a question which we have often addressed to ourselves, namely,— under what aspect must the very varied changes which the solar light experiences on the surface of our planet be presented to the inhabitants of Mars and Venus? A more attentive observation of the ashen-gray light of the moon, which appears to be principally produced by the reflection of the more or less luminous face of the earth, may perhaps one day provide us with an answer to our question.

Before quitting this subject, let us remember that both snow and frost are of great utility to the husbandman. The latter, by expanding the humidity with which the hard clods are penetrated, crumbles them into powder, and renders stiff land porous, friable, and mellow. It also clears the soil from the plague of insect life, which, if it increased without so powerful a check, would probably prove a terrible injury to the crops. Moreover, it so hardens in winter the moist soft ground as to permit of the necessary field operations being carried on. Snow, as Dr Child remarks,* is even more useful. It covers up the tender plants with a thick mantle, which defends them against the attacks of excessive cold. "God giveth snow like wool," and for somewhat the same purposes as wool. The mantle which so closely wraps about the gaunt limbs of the winter-stricken earth neither allows the internal heat to escape

* Chaplin Child, "Beredici," p. 171.

nor the external cold to enter in. It has been found that the inner surface of the snow seldom falls much below 32° F., although the temperature of the external air may be many degrees under the freezing-point; and it is known that this amount of cold can be endured by the crops without injury, so long as their covering protects them from the raking influence of the wind. In climates where the winter's cold is longer and more intense than in England, the protective influence of snow is much more plainly shown. Where it lies long and deep, it opens out routes that were impracticable in summer on account of their ruggedness, and prepares a smooth path for the sledge, or for the "lumberer," over which the largest trunks of the forest may be carried with ease to the river or canal.

In the polar regions (we quote from Dr Child) snow supplies the ever-ready material out of which the Esquimaux construct their houses, and hardy explorers extemporise the huts in which they find shelter when absent from their ships on distant expeditions. Nor are the ships themselves considered "snug winter quarters" until their sides have been banked up in walls of snow, and the roof raised over the deck has been thickly covered with it. Snow huts are warmer than might have been expected. If built upon ice over the sea, their temperature is sensibly influenced by the heat of the unfrozen water below, which is said seldom to fall much under 40° F. in any part of the ocean. Even where the external temperature has sunk to 20° or 30° below zero, sufficient warmth is produced in a snow hut by the huddling together of three or four persons within it. When Dr Elisha Kane, the American explorer, passed a cold arctic winter's night in a hut beyond Smith's Sound, the temperature produced by its complement of lodgers, and two or three oil lamps, reached 90° F.; so that he was compelled by the heat to follow the example of the rest of the party, and partially to divest himself of his clothing. Yet in lat. 79° N., Dr Kane marked a temperature of 75° below zero in the month of February. No fluid could resist it. Even chloric ether became solid, and the air was pungent and acrid in respiration.

RED SNOW.

As if it had been ordained that there should be nothing absolute in nature, snow itself, the very type of whiteness, sometimes exhibits the most curious colouring. Who, for instance, has not heard tell of *red snow?* Its existence was even known to Pliny, the great Roman naturalist, and he attributed it to a dust with which the snow became covered after it had lain several days on the ground. "Snow itself," he says,* "reddens with old age" (*Ipsa nix vetustate rubescit*).

Benedict de Saussure was the first who described red snow like a naturalist.† He observed it on the occasion of his ascent of Mont Breven, near Chamounix, in 1760; and was greatly astonished at seeing the snow tinted in various

---

* Pliny, "Historia Naturalis," Book xi.

† De Saussure, "Voyage dans les Alpes," Book iii., p. 45 (ed. 1803, Neufchâtel).

places of an extremely vivid red. "In the middle of each patch," he says, "was the greatest intensity of colour, and the middle, moreover, was of a lower level than the edges. On examining this red snow closely, I saw that its colour depended upon a fine powder which mingled with it, and which penetrated to a depth of two or three inches. This powder could not have descended from the summit of the mountain, since it was found in localities isolated and even remote from the rocks; nor did it seem to have been deposited by the winds, since it did not lie in drifts. One would have said that it was a production of the snow itself, a residuum of its thaw. . . . What at first suggested this opinion was the fact that the colour, extremely weak on the edges of each concave patch, gradually grew deeper as it approached the bottom, where the trickling water had carried down a greater quantity of residuum."

The learned Swiss naturalist found this red snow on many other mountains, and during a certain period of thaw, subjected it to various experiments, which led him to the conclusion that it was a vegetable matter, "a dust, or pollen, of the stamens of plants." Slightly odorous, it exhaled, during combustion, a scent not unlike that of sealing-wax.

Ramond met with red snow in the Pyrenees, at an elevation of 7800 feet. He discovered in it, when burnt on incandescent coals, the odour of opium or of chicory. He supposed that the little deep red lamellæ which coloured the snow were mica, and looked upon the mica as a product of the decomposition of the rocks by the action of the sun and breezes of spring. But this opinion was overthrown by Captain Ross, who, in 1819, found red snow in Baffin's Bay (lat. 85° 54′ N.), to a depth of thirteen feet, over a soil perfectly free from mica. Other explorers affirm that in those regions they have never met with the red snow more than three to four inches deep. Captain Parry, in his Polar voyage, found this coloured snow principally in the track of his sledges; and, agreeing with Sir John Ross, he supposed it to derive its redness from the presence of a kind of mushroom, of the genus *Uredo*, to which Bauer has given the name of *Uredo nivalis.** According to experiments made by Bauer on specimens brought from the Polar regions, these tiny mushrooms are, on the average, a fiftieth of a millimètre in diameter; they develop themselves like vegetables; the youngest are sometimes colourless; when entirely freed from snow, they grow black under the influence of an intense cold, without losing their germinative faculty, and give birth, under the influence of a higher temperature, to a green matter.

Let us continue to examine the difference of opinion between naturalists.

De Candolle declared the red snow of the polar regions to be identical with that of the Alps, after having carefully compared the two. But he saw in it a genus of cryptogams, differing from the genus *Uredo*.† Robert Brown asserted that

---

* Annales de Chimie et de Physique, vol. xxvii., p. 134.
† Philosophical Transactions, 1820, vol. ii., p. 165.

it was a kind of alga, allied to the *Tremella cruenta*. Azara was of this same opinion, except that, instead of a tremella, he recognised in it an alga of the genus *Protococcus*, which he called *Protococcus kermesinus*, because its colour resembled that of the kermes, or cochineal.

In the opinion of the observers whom we have cited, the colouring corpuscles of the snow belong to the vegetable kingdom. This opinion was supported by numerous adherents, and soon acquired so great an authority, that, in an assembly of naturalists at Lausanne, De Candolle overwhelmed with sarcasm a communication from Lamont, Prior of the Hospice of St Bernard, on the "animality of red snow." And yet this last hypothesis was not so rash as might have been supposed; for Dr Scoresby, to whom we owe a profound study on the crystalline forms of snow, had already attributed to an animal matter the colouring of the snow and polar ice.

Now-a-days, however, it may be regarded as finally settled that this phenomenon is due to the immense aggregation of minute plants belonging to the species called *Protococcus nivalis*;* so called in allusion to the extreme simplicity of its organisation, and the peculiar nature of its habitat. If we place a portion of the snow coloured with this plant upon a piece of white paper, says Mr Macmillan,† and allow it to melt and evaporate, we find a residuum of granules just sufficient

---

* It is but fair to add, however, that Vogt and others contend for the animal origin of this substance, and regard the *Protococcus nivalis* as simply a development of the infusoria, *Disceræa nivalis*.

† Macmillan, "Footnotes from the Page of Nature," pp. 141-143.

to give a faint crimson tinge to the paper. Placed under the microscope, these granules resolve themselves into spherical purple cells, from the 1/1000th to the 1/3000th part of an inch in diameter. Each of these cells has an opening, surrounded by serrated or indented lines, whose smallest diameter does not exceed the 1/5000th part of an inch! When perfect, the plant is not unlike a red-currant berry; as it decays, the red colouring matter fades into a deep orange, and the deep orange changes into a dull brown. The thickness of the wall of the cell does not exceed the 1/50000th part of an inch! Each cell may be considered a distinct individual plant, since it is perfectly independent of others with which it may be aggregated, and performs for and by itself all the functions of growth and reproduction, having a containing membrane which absorbs liquids and gases from the surrounding matrix or elements, a contained fluid of peculiar character, formed out of these materials, and a number of excessively minute granules, equivalent to spores, or, as some would say, to cellular buds, which are to become the genus of new plants. There is something, adds Mr Macmillan, extremely mysterious in the performance of these widely different functions, by an organism

FIG. 8.—Protococcus nivalis.

which appears so excessively simple. That one and the same primitive cell should thus minister equally to absorption, nutrition, and reproduction, is an extraordinary illustration of the fact, that the smallest and simplest organised object is, in itself, and for the part it was created to perform in the operations of nature, as admirably adapted as the largest and most complicated.

The Eternal Snows.

The epithet "eternal" or "perpetual," applied to snow, would appear to savour of the ambitious, if not of the profane. Can we say of anything which belongs to earth that it is "eternal?" Assuredly not. The earth has not always worn the aspect which it now wears, and, at a period not far distant from its origin, could not in any region have been covered with snow. Now, whatever has had a beginning cannot be eternal. Many authors have, for this reason, substituted for the word *eternal* the word *perpetual*. But the latter is equally inapplicable. Who will venture to affirm that our globe or its system will endure *perpetually* ?

This difficulty, however, is one which need not particularly embarrass us. We have been long accustomed to look upon language as a simple mask, or, at least, as a very dubious interpreter of thought. And we shall, therefore, continue to use indifferently the words "eternal" and "perpetual."

Let us suppose that two travellers set out from the equator, that plane of separation between the northern and the southern hemispheres. Let us also suppose that each proceeds in a diametrically opposite direction to the other, pursuing his route along one of those meridian lines which divide the earth into longitudinal portions, like the slices of a melon (to compare great things with small). The following will be their climatic stages:—

At first the two travellers will each traverse a moiety of the *torrid zone*, limited below and above the Equator by two parallel circles,—in the northern hemisphere by the Tropic of Cancer, and in the southern by the Tropic of Capricorn. Do not let these appellations alarm you: they show, once more, the narrow connexity of the heaven with the earth; *tropic*, coming from the Greek τροπή, signifies a return—the sun returns from his apparent excursions, after having passed from the tropics to the zenith. For these circles form the extreme limits of the sun's march towards the north and towards the south: they are the two solstices—the summer solstice, when the sun enters the zodiacal sign of Cancer, and the winter solstice, when it enters the sign of Capricorn. The torrid zone is the only one which is thus divided into two portions by the Equinoctial, and which the sun passes twice a year to the zenith, that is, to the point directly above the heads of the inhabitants.

After having crossed the tropics, one of our two travellers will enter the *North Temperate Zone*, bounded by the Arctic Polar Circle,—the other, the *South Temperate Zone*, bounded by the Antarctic Polar Circle. Having passed the polar circles, they will find themselves speedily arrested by ice and snow which never melt—by eternal ice and snow. These inhospitable regions compose the two *frigid zones*, which cover, like two immense hoods (forming the 0·82 parts of the terrestrial surface), the one, the northern hemisphere, the other, the southern.*

In their progress through these various climates, our two travellers will arrive at a very curious comparative result,— that the southern hemisphere is colder than the northern. This difference becomes especially recognisable below the 50th degree of south latitude; so that, after passing the Antarctic Circle, the ice opposes the voyager's course with nearly insurmountable obstacles; while, in the northern hemisphere, the whaler frequently penetrates to Spitzbergen, situated much nearer to the Pole than to the Polar Circle. This is a general fact; we confine ourselves to putting it forward.

Let us now suppose that our two travellers, always ready to compare the results of their inquiries, accomplish the ascent of a very lofty mountain situated under the Equator, such as Chimborazo. In the course of their ascent, they will traverse the same climates and the same zones which had marked the stages of their journey from the Equator to the Poles: at their starting-point they will find themselves in the Torrid zone, then will come the Temperate and the Frigid zones, the latter rendered inaccessible by glaciers and eternal snows. These vertical zones of the mountain are characterised by vegetables and animals whose types are found in the corresponding horizontal zones of the terrestrial surface. But what is most

---

* The two temperate zones together represent perceptibly the half, or 0.520, and the torrid zone, two-fifths, or, more exactly, 0.398, of the terrestrial surface.

remarkable is, that there exists between the northern and the southern slopes of the mountain the same difference as between the southern and northern hemispheres: the line of the eternal snows descends much lower on the northern than on the southern slope, in the same manner as, in the southern hemisphere, the polar ice advances much nearer the Equator than in the northern.

Such is the general view-point which we must adopt for the clearer comprehension of the details of observation. Of course, when speaking of the limit of the eternal snows, we refer only to the *lower* limit, that is to say, to the greatest elevation attained by the snow-line in the course of a single year. As for the *upper* limit, it entirely escapes us; for the summits of the loftiest mountains do not reach the atmospheric strata which, by virtue of their refraction, cannot contain any vesicular, aqueous, or condensable vapour.

The line of eternal snow which, *at the poles*, is found on the level of the ground, gradually rises as we approach the torrid zone, where it attains its maximum of elevation, from 13,000 to 17,000 feet. This phenomenon does not exclusively depend upon the geographical latitude, nor on the mean annual temperature of the locality: it is the result of an aggregate of diverse circumstances which we have not the space here to enumerate and discuss. We shall content ourselves with placing before the reader a table which will show the remarkable differences existing in the height of the perpetual snow-line in various places.

The Line of Perpetual Snow.

Latitude.	Place.	Height of Snow-Line.
Degs.		
79 N.	Spitzbergen	0
71	Mageroe (Norway)	2,350
70 to 60	Norway (Interior of)	3,500 to 5,100
65	Iceland	3,050
54	Oonalashka (W. America)	3,510
50	Altai Mountains	7,034
45	Alps, The, N. declivity	8,885
45	Do., S. declivity	9,150
43	The Caucasus	11,863
43	The Pyrenees	9,000
40	Mount Ararat	14,150
36	Karakorum, N. side	17,500
36	Do., S. side	19,300
36	Kuen-luen, N. side	15,000
35	Do., S. side	15,680
29	Himalaya, N. side	19,560
28	Do., N. side	15,500
17	Cordilleras of Mexico	14,650
13	Ethiopian Mountains	14,075
1 S.	Andes, in Quito	15,680
16	Do., in Bolivia, E.	15,800
18	Do., in Bolivia, W.	18,400
33	Do., in Chili	14,600
43	Do., in Patagonia	6,300
54	Strait of Magelhaens	3,700

The Inhabitants of the Eternal Snows.

If men have the faculty of living under all climates, they make use of that faculty, as we know, with extreme reserve. They have never permanently inhabited the polar regions and the perpetually snowy summits of the mountains: it is only at intervals that a few pioneers have temporarily ventured thither. Starting from this fact, it was long believed that the zone of eternal snows was not inhabited by any living being. Even men of science admitted, as an article of faith, that where man could not fix his residence no animal could live. They made, however, a concession with respect to vegetables, and particularly as regarded the lichens and the mosses.

Well, observation and research conjointly, have erased this article of faith from the scientific code. It has been demonstrated that the icy regions, which man visits only at rare intervals, and where he sojourns but for a time, are the home of a certain number of animal species, more or less allied to the human species. The scientific exploration of these regions dates only from our own time. Spitzbergen, and the summit of the Alps,—such are our points of comparison.

It is difficult to conceive of anything more interesting than the historical exposition of the limited *Fauna glacialis*. First, let us take the discovery, comparatively recent, of a small rodent of the mouse order.

The Arvicola Leucurus, or Arctic Vole.

On the 8th of January 1832, a Swiss naturalist, M. Hugi, started from Soleure to study the winter condition of the classic glacier of Grindelwald. The undertaking was in many respects a difficult one; the sides of the Mettenberg, bordering on this glacier, were covered with an uniform stratum of hardened snow; a pathway had to be cut out with the pickaxe. M. Hugi and his companions did not arrive at the Stierreg until towards evening.

A goatherd lives there during the summer. They sought

around and about for his little cabin, but, on the uniform white

FIG. 9.—Among the Alps.

carpet of snow which covered everything, no sign of it could be detected. At length they came upon a little mound, which they immediately proceeded to excavate; and late in the night they discovered the roof of the hut. They then redoubled their efforts to sweep away the snow obstructing the door. On opening it a score of mice emerged from the cabin, some of which they killed.

For a picture of the poor victims we are indebted to M. Hugi. "These little rodents are of a yellowish gray, and very slender; from the head to the tip of the tail they measure about nine inches. The hind paws are of a length wholly disproportionate to the fore paws. The tail and ears are naked; their transparency is remarkable. . . . This animal," adds M. Hugi, "appeared to me completely unknown, and I do not remember to have seen it in any zoological collection."

After determining its genus and species, the intrepid explorer of the Alps was entitled to have given it a name; but this honour escaped him.

The same little rodent has since been found in many other parts of the Alps; notably among the rocks of the Grands-Mulets, some 12,500 feet above the sea-level.

FIG. 10.—The Arctic Vole.

Desirous of comparing the climate of Spitzbergen with that of the summit of the Alps, M. Martins established himself, in 1841, with his friend, A. Bravais, on the Faulhorn. "While," he says, "we were engaged in our experiments, we often perceived a little animal passing swiftly by us, and stealthily gliding into its burrow. We remarked that it was also found in the *auberge*, or inn, and that it fed upon Alpine plants. At the first glance, its resemblance to the common mouse led us to think that this inconvenient guest had followed man into his abode on the Faulhorn, as it had formerly crossed the seas on board ship. But a more attentive examination showed me that, far from being a mouse, it was a species of vole, which had hitherto escaped the researches of naturalists. I designated it by the name of the snow-vole, Arvicola nivalis."

It was the same animal which M. Hugi had discovered nine years before. The ice was broken, and names, both generic and specific, afterwards fell like hail on the body of our poor little rodent. Some would have had it called—

   Hypudæus alpinus.   Hypudæus nivicola.
   Hypudæus petrophilus. Hypudæus Hugei.

Others, and fewer in number, preferred the designation of "White-tailed Vole,"—

     Arvicola leucurus.

Others again, "the Lebrun vole,"—

     Arvicola Lebrunii.

Which of all these names shall prevail? We cannot say, and it matters very little to us. Perhaps the nomenclators may in time agree among themselves upon the appellation of the genus. However this may be, we know—and it is an important fact—that a mammal exists at altitudes where men could not live, and that he is found in the Alps, above even the lower limit of the perpetual snows. Is it the only mammal which can live at such a height?

The Marmot.

Who, in the wide world of London, where so many waifs and strays are drifting with the great current of human life, has not observed the Savoyard wanderer with his dancing marmot? If the man did not attract our notice, his curious companion would. In form he belongs partly to the bear, and partly to the rat. Naturalists have, therefore, expressly created for him the genus *Arctomys*,—a Latinised Greek name, signifying "The Bear-Rat."

In fact, the marmot resembles Harlequin's cloak, or rather, if it be permissible to compare little things with great, the Austrian Empire,—a composite of territories and races; and Buffon has described him very aptly. He has, he says, the nose, the lips, and the shape of the head of the hare; the hair and nails of the badger; the teeth of the beaver; the cat's whiskers; the eyes of the dormouse; the feet of the bear, with a short tail and truncated ears.

Add to this that the marmot—he is a little larger than a rabbit—is omnivorous like man and the bear, with whom he shares his aptitude for dance and sport. While he eats any and everything, he nevertheless prefers vegetable food to all other kinds; and with his orange-coloured incisors gnaws the bark of shrubs. He rarely drinks, but when he does drink takes a hearty draught; is particularly fond of milk; drinks it by raising his head at each mouthful, like a hen, and giving utterance to an audible murmur of contentment, just as if he were reciting his *Benedicite*. Will it be in allusion to this characteristic that the common French phrase has originated, *Marmotter des prières ?* *

During the summer the marmots inhabit the snowy summits of the Alps. At the beginning of autumn they descend to a lower level, for the purpose of excavating the burrows in which they pass the winter, completely benumbed by the frost. This is the time when the hunters easily capture them; they have nothing to do but to dig (*creuser* is the technical word); and frequently they are found as many as ten or twelve in the same burrow, rolled up like balls, and buried in a litter of hay. Their sleep, says De Saussure, is so profound, that the hunter deposits them in his sack and carries them away without awakening them. The Chamounix hunters, he adds, have already entirely expelled or destroyed the goats formerly so abundant on their mountains; and it is probable that, in less than a century, we shall see neither chamois nor marmots.

This prophecy of De Saussure's is on the point of being realised. Still, even at the present day, marmots are not very rare in the Valais and the canton of Ticino (du Tissin), where they are called *Mure montane* (mountain rats); a phrase from which is derived, without doubt, the appellation *marmot*. They prefer as their abode the stony islets which rise here and there in the midst of the rocks. The ears of travellers who venture into the barrenest recesses of the Alps of the Bernese Oberland are sometimes struck by a very sharp

* That is, muttering, marmot-wise, one's prayers.

whistling, for which, at first, they find it difficult to account. It is the young marmot's cry of alarm; for the old appear to be deprived of this strident faculty.

For a considerable period only a single species was known —the marmot properly so called (*Arctomys marmotta*, Gmelin); but four others must now be added :—1st, The marmot of the Caucasus (*Arctomys musicus*), still imperfectly known; 2d, The marmot of Canada (*Arctomys empetra*), who clambers up the trees like a cat, and is distributed throughout all North America, particularly in Hudson's Bay, and Alaska, on the north-west coast; 3d, The *Arctomys monax*, who appears to be peculiar to Maryland; 4th, The Russian marmot (*Arctomys citillus*), of the size of a field-mouse, and of a brown colour, spotted with white; 5th, The marmot of Siberia (*Arctomys bobac*), smaller than the common species, of a yellow gray, and building vast burrows shaped like a funnel.

Will the reader permit us an allusion, in passing, to a question which we do not see discussed in books of natural history? Formerly among the treasures of ancient druggists figured a kind of panacea, called "Graetz's balls." What were these "Graetz's balls," at one time esteemed as a universal medicine, but no longer included in our pharmacopeia?

This was their origin :—The subterranean dwellings which certain species of marmots construct with so much skill, are each composed of two galleries, which unite together like the arms of a Y, and terminate in a *cul-de-sac.* There are found the globules of clay known as "Graetz's balls." They are an industrial product of our rodents, as M. Oscar Schmidt established in 1866, by close observation of the *Arctomys bobac* of the Zoological Garden of Vienna. The marmot creates these balls by scratching up the earth, and appears to amuse himself—a child's amusement!—by rolling them to and fro in his galleries.

The Chamois.

> "Even so
> This way the chamois leapt." *

Must we omit this graceful ruminant from the number of mammals inhabiting the eternal snows? No; for it is not of his own will that the chamois has taken refuge upon the snowy peaks of the Alps. If we meet with him there, it is because he seeks to shelter himself from the destructive instincts of man.

The chamois is one of those animal species which, before a century, perhaps, will have disappeared; his bones will then figure in the palæontological museums by the side of the skeletons of extinct species. There, too, will be displayed the famous *chamois balls*, each of the size of a nut, covered with a shining substance resembling leather, of an agreeable odour, and seeming to be a morbid dejection, composed of roots and other undigested matter. These balls, the *bezoars* of the old physicians, were regarded as a remedy against every ill the human flesh is heir to; it was even professed that they rendered soldiers invulnerable, and were a better defence against bullets than the finest armour ever wrought by the smiths of Milan. How precious a remedy for this epoch of civilisation, when man—is he much wiser than his supposed

* Byron.

progenitor, the ape?—seeks to replace the cholera and the pestilence by the most terrific engines of destruction!

The birds inhabiting the inhospitable region of the snows are more numerous than the mammals. Let us briefly refer to a few of the more important.

The Eagle and the Wren.

In speaking of the eagle, Tennyson's noble lines to that "imperial bird" will occur to every reader, from the force and clearness of the picture which they present:—

> "He clasps the crag with hookèd hands;
> Close to the sun in lonely lands,
> Ringed with the azure world, he stands.
>
> "The wrinkled sea beneath him crawls;
> He watches from his mountain walls,
> And like a thunderbolt he falls."

The affection of the eagle for his "mountain walls" may be easily understood. This giant bird, with his carnivorous instincts, is endowed with a remarkable tenacity of life, and can exist in habitats wholly inaccessible to man. But it is strange that a bird, which is as common a type of humility as the eagle is of ambition, and which we almost always cite as a contrast to the eagle—we mean, the delicate little wren—should also be found among the snow and ice, the silence and solitude, of the loftiest mountain regions.

To study the flight of the eagle, we should repair to Alpine highlands. When he has reached a certain altitude of the atmosphere, the royal bird descends obliquely, as upon an inclined plane, with a rush and a din of wings, and at a speed of upwards of thirty-six yards per second. We assured ourselves of the accuracy of this fact during an ascent of Mount Hochkœrpf, in the canton of Glaris. The bird traversed in six minutes a space of 40,000 Swiss feet, which is equal to about forty yards per second. This result agrees, on the whole, with the observations of a traveller, who wrote in the *Nouvelle Gazette de Zurich*, on the 26th of August 1863 :—

"A society of tourists set out from Corri to climb the Stützerhorn, which is 8400 feet in height. From the summit they perceived an enormous eagle, which, having taken his flight from Calanda, beyond the Rhine, directed his course towards the Stützerhorn, for the purpose of halting, after a certain inflection, on the side of the Rothhorn."

The duration of the flight was five minutes; the interval between the starting-point and the point of arrival, two French leagues and a half. In three hundred seconds, therefore, the eagle must have traversed a space of 3000 Swiss feet, which is equal to a speed of forty-five yards per second. Hence, the swiftness of the eagle's flight is nearly equal to the velocity of sound.

One of the most admirable descriptions of the habits of this bird with which we are acquainted, is furnished by the well-known naturalist Macgillivray :—

"There he stands"—on his lonely crag—"nearly erect, with his tail depressed, his large wings half raised by his side, his neck stretched out, and his eye glistening as he glances around. Like other robbers of the desert, he has a noble aspect, an imperative mien, a look of proud defiance; but his nobility has a cast of clownishness, and his falconship a vulturine tinge. Still he is a noble bird, powerful, independent, proud, and ferocious, regardless of the weal or woe of others, and intent only on the gratification of his own appetite; without generosity, without honour; bold against the defenceless, but ever ready to sneak from danger. Such is his nobility, about which men have so raved.

"Suddenly he raises his wings, for he has heard the whistle of the shepherd on the crag, and bounding forward, he springs into the air. Hardly do those vigorous flaps serve at first to prevent his descent, but now curving upwards, he glides majestically along. As he passes the corner of that buttressed and battlemented crag, forth rush two ravens from their nest, croaking fiercely. While one flies above him, the other steals beneath, and they essay to strike him, but dare not, for they have an instinctive knowledge of the power of his grasp; and, after following him a little way, they return to their home, vainly exulting in the thought of having driven him from their neighbourhood.

"But on a far journey, he advances in a direct line, flapping his great wings at regular intervals, then shooting along without seeming to move them. In ten minutes he has progressed three miles, although he is in no haste, and now disappears behind the shoulder of the hill. But we may follow him in imagination, for his habits being well known to us, we may be allowed the ornithological license of tracing them in continuance."

Homeward bound,—Mr Macgillivray continues,—after having supplied his own wants, he knows that his young must be provided with food. Therefore he sweeps across the moor, at a height of two or three hundred feet, bending his course to either side, his wings wide-spread, his neck and feet retracted,

FIG. 11.—The Eagle's Habitat.

now beating the air, and again sailing smoothly along. Suddenly he halts, poises himself for a moment, stoops, but recovers himself without touching the ground. The object of his regards, a golden plover, which he spied on her nest, has contrived to elude him, and he does not care to pursue her. Now, then, he ascends a little, wheels in short abrupt curves, presently rushes down headlong, assumes the horizontal position when close to the ground, prevents himself from being dashed against it by expanding his wings and tail, thrusts forth his talons, and grasping a poor, terrified ptarmigan that sat cowering among the gray lichens, squeezes it to death, raises his head exultingly, utters a clear, shrill cry, and, rising from the ground, pursues his journey.

As he passes a tall cliff overhanging a silent lake, he is attacked by a fierce peregrine falcon, which darts and plunges at him, as if resolved to deprive him of his booty, or drive him headlong to the ground. A more formidable foe is this than the raven; and the eagle, with a scream and a yelp, throws himself into postures of defence, until, at length, the hawk, perceiving that the tyrant has no intention of plundering his nest, leaves him to pursue his course without further molestation. Over dense woods, and green fields, and scattered hamlets, the eagle speeds; and now he enters the long river-valley, near whose upper end, cradled in mist, rises the rock of his eyrie. About a mile from it he meets his mate, who has been abroad on a similar design, and is returning with a white hare as her spoil. With loud strident cries they congratulate each other, cries that alarm the drowsy shepherd on the green strath below, who, remembering the lambs carried off in springtime, discharges at them a volley of maledictions.

Their nest is of considerable size, but rudely constructed;

a pile of twigs and heather and dead sticks, somewhat hollow in the middle, where lies a thin deposit of wool and feathers. Here repose the eaglets, two in number, and clothed in soft white down.

Independently of the species which, like the *Pandion haliætus*, and the *Aquila nævia*, inhabit the lower regions, the eagles which visit the Alps are very remarkable. Thus, the

FIG. 12.—The Lammergeier.

species of *Gypaëtos*, which the inhabitants designate under the name of the *Lammergeier*, or " Lamb-slayer," is the European condor. The spread of his wings is about ten feet; he weighs from eighteen to twenty-four pounds, and can easily carry off in his talons kids, lambs, and even children.

The *Steinadler*, which, like the preceding, belongs to the inaccessible mountains of the cantons of Glaris, Schwyz, the Grisons, Appenzell, and Berne, would seem to be a variety or sub-species of the *Aquila imperialis*. The inhabitants of Eblingen, a village on the borders of the Lake of Brienz, hunt him vigorously. Finally, some eagles there are which only sojourn in the Alps temporarily; they appear to be astray; such are—

The *Circætus leucopsis*, which has a particular affection for serpent-haunted districts;

The *Haliætus leucocephala*, with head and tail of a milky white, belonging to the north of Europe and America; and

The *Neophron percnopterus*, or Egyptian eagle, of carrion-like odour, which is sometimes met with in the neighbourhood of Geneva.

The tawny-headed vulture (*Vultur fulvus*), and the ashy vulture (*Vultur cinereus*), with gray-brown mouth, and a brownish collar round his bare neck, are extremely rare in Switzerland.

But we now take leave of the eagle, and turn our attention to the lowly wren, whose charming but simple music has been described in charming but simple verse by Bishop Mant:—

> "The quick note of the russet wren,
> Familiar to the haunts of men;

> He quits in hollowed wall his bower,
> And through the winter's gloomy hour
> Sings cheerily; nor yet hath lost
> His blitheness, chilled by pinching frost,
> Nor yet is forced for warmth to cleave
> To caverned nook or straw-built cave,—
> Sing, gentle bird! sing on, designed
> A lesson for our anxious kind,
> That we, like thee, with hearts content"——

The wren here referred to is a British species, the common wren, or *Troglodytes vulgaris*, one of the smallest of our British songsters; a restless, lively bird, which twitters about the hedgerows in summer, and about the garden and shrubbery in winter, and chanting his mellow song even under the gloomy sky of December. Allied to this familiar bird is the Gold-crowned Knight,* or *Sylvia regulus*, which is found in the Alpine deserts at an elevation of 9000 to 10,000 feet. Like our own Jenny Wren, he has a very fine, slender bill, which denotes his insectivorous propensities. He is easily known by the little crest of silky feathers which he wears on his head, like a diadem, and also by his peculiar cry of *souci-i-i-i*.

FIG. 13.—The Wren.

Our crowned knight is very partial to the society of the tits, and, like them, he is easily caught with bird-

---

* Also called the Golden-crested Wren.

lime. He is so fond of the company of other birds, that, when he finds himself alone, he becomes disquieted; his prolonged tiny chirp grows plaintive; and he flies to and fro in quest of comrades. He may be regarded as a trustworthy barometer, for, prior to rainy weather, his song is very loud and incessant. Devoted to the pursuit of insects or their larvæ, he seems to pay no attention to the passer-by; he flutters vivaciously from branch to branch, and puts himself in all imaginable positions, sometimes with his head upwards, sometimes with his head downwards. We have often watched, with extreme gratification, the acrobatic tricks of our Lilliputian gymnast. Occasionally, before he perches, you will see him, in a frenzy of indecision, rapidly agitating his wings, and revolving them like a wheel. If you look at him, while thus engaged, *against* the light, you will think you see a tiny, ethereal, diaphanous spinning-top. After "assisting" at such a spectacle, which the first wood will furnish, you will not be indisposed to admit with us, that the bird designated by the Greeks τροχίλος, or "little wheel," and whose identity has so often been discussed, was, in reality, our golden-crested knight. Moreover, he is a true cosmopolite, in every acceptation of the word. Not only does he never quit us, not only does he remain faithful to us throughout the year, but we meet with him over all Europe. He is also found in Asia, and even, it is said, in America, from the West Indies to Canada. His flight being very short, it is supposed that he passed from one hemisphere to another by way of Behring's Strait. It is certain, at all events, that he discovered the New World before Christopher Columbus.

During the severe cold of the winter of 1867-8, we saw our knight—a very rare circumstance—haunt the vicinity of our houses, though he prefers the green shade of the forests, and especially of the forests of pine and fir. He who has seen him pecking at the bark and leaves of these trees, while the ground was covered with snow, and during a frost of 10° below zero (C.), will feel no astonishment at meeting him upon the snowy summits of the Alps.

Yet this pet bird of ours, this Lilliputian warbler, does not weigh more, with all his feathers, than a quarter of an ounce, or the two thousandth part of an eagle. Away with the hunter who would attempt such tiny game! A bird so small that he glides through the meshes of a net,—so delicate, that if you would not irreparably injure your "specimen," you must shoot him with a few grains of finest shot,—a bird of such frail appearance, withstanding all climates, and distributed over the entire surface of the globe,—here is a subject worthy the meditation of man, who pretends to be the "lord of creation!"

THE SNOW-BUNTING—(*Emberiza plectrophanes nivalis*).

This is the snow-lark-bunting of Macgillivray, and a species of the genus *Plectrophanes*. In Scotland, he is frequently called the Snow-flake, and, in other parts of Great Britain, the Snow or Oat-fowl. His weight does not exceed an ounce and a half. His bill and legs are black; his forehead and crown white, with an admixture of black on the hind part of the head; black are the back and sides, but each wing is marked by a broad belt of white; the quill feathers are

FIG. 14.—The Crests of Inaccessible Rocks.

black, with white bases; the secondaries are white, with black spots on the interior webs.

The snow-bunting's favourite localities, where he loves to

build his nest, are the crests of inaccessible rocks, surrounded by vast fields of snow, in whose midst the sun and tempest have created a few oases. The most he does is to approach the hospices of Monts St Bernard and St Gothard, and construct his nest under the eaves of their roofs. This nest, made of long blades of grass, is lined internally with hair and the feathers of poultry. At the beginning of May, the female lays six eggs of a snowy whiteness, and the male assists in hatching them, and bringing up the young. The bill of the latter is, at first, of a bright yellow, which turns black, like that of the parents, as they grow older.

The snow-bunting rarely descends into the wooded region. Of a very sprightly disposition, he spends nearly all his life in the midst of the snows and the ice. His song somewhat resembles that of the finch, which he also resembles in size and social instincts; for he may frequently be seen in numerous bands hovering above the highest mountains.

The snow-bunting is also met with in the northern districts of Asia and America.

The Red-Billed Crow—(*Corvus pyrrhocorax*).

The familiar cry of this bird, who resembles the thrush,— the *krapp-krapp* of the red-billed crow (*la corneille des nieges*),— agreeably falls on the ear of the traveller, when wandering through solitudes devoid of any other living being. By their cries and their presence these birds animate the denuded rocks which rise like promontories in seas of ice. They are easily distinguished from other species by their coral-red bills;

whence their name of *Pyrrhocorax*. They nest in troops in the crevices of the most inaccessible rocks, and propagate there from generation to generation. Their presence is indicated by enormous heaps of ordure, veritable guano, which might well be used for manure. Their abrupt ascents and strident cries are signs of bad weather, which the mountaineer knows how to profit by.

If caught when young, these birds are easily domesticated. M. Tschudi, in his "Life among the Alps," relates the history of one who had been tamed. He would himself go in quest of the bread, cheese, and fruits which composed his repast; then, holding in his claws the prize he had coveted, devour it with avidity. What remained of his meal he carefully concealed in paper, and would gallantly defend the hidden treasure against whomsoever dared to approach it, against dogs as well as man. Fire had a singular attraction for him; he would extract from a lamp the burning wick, and swallow it without sustaining any injury; he would swallow even the *débris* of the charcoal as he fluttered about the chimney. He showed an excessive joy at the sight of smoke, and loudly clapped his wings. Whenever he caught sight of any burning coal, he did not fail to pick up immediately all the paper, rags, or twigs he could lay his claws on; these he would place in the stove, and amuse himself by watching the smoke they gave forth. If a stranger entered the room, he gave vent to the most deafening cries, though he was exceedingly gentle and familiar towards persons with whom he was acquainted. His friends and favourites he distinguished in a peculiar manner; he ran in front of them, displayed his joy by expanding his wings, and alternately perched himself on their hands, their head, their shoulders, eyeing them all over, and bending his head as if to kiss them. Every morning he entered his master's bedroom, called him by his name, posted himself on his pillow, and waited tranquilly until he awoke; then he expressed his satisfaction by all kinds of gestures and noises.

Reptiles.

Close to the line of perpetual snow a black variety of vipers has been met with; but none of the serpent race ever *cross* that line.

The only reptile found within the boundaries of the snowy region is a kind of lizard (*Zootoca pyrrhogastra*), the only one, perhaps, of all the vertebrata which could live at an elevation above the sea-level of more than 9500 feet, buried in the snow for upwards of ten months.

During the few bright summer weeks, he feeds upon some rare insects and spiders.

The frigid zone is so far the natural habitat of these lizards, that they would rather die of hunger than live in the more genial regions to which men have wished to transplant them. In length they nearly equal our common lizards, but they are not quite so big; their back is of a chesnut brown, marked with black streaks and dots; the throat is bluish; the belly of the male is of a greenish blue, spotted with black, while that of the female is of so lively a red as to have suggested the name of the species, *Pyrrhogastra*; just as the name of the genus is derived from the circumstance that the young are hatched in the mother's belly, and are born alive like the young of a mammal. This statement, too, holds good with respect to the viper, which also endures the cold of elevated regions.

INFERIOR ANIMALS.

Our information is still very incomplete so far as relates to the molluscs, the arachnida, and the insects which inhabit the frigid zone. The Alpine snail (*Helix Alpicola*), so remarkable for its transparency, appears to be the sole mollusc which, in certain localities, attains to an elevation of 7000 feet. It is, however, surpassed by the earthworm, which is not only distributed over the surface of every country, but ascends to the snowy summits of the loftiest mountains. Few animals have their geographical distribution so extended both horizontally and vertically; and only some species of spiders and millepeds keep company with the earthworm.

Among the other inhabitants of the snows have also been observed a dozen species of butterflies,—nearly all diurnal,—for the phalænæ (?), or nocturnal Lepidoptera, appear to be much more sensible to the cold. M. Agassiz saw the "Little Vulcan" (*Vanessa urticæ*) fluttering in the snowy desert which borders on the glacier of Aar, as if it were completely in its element. The wings of the majority of these butterflies are sombre-coloured; their caterpillars live upon the auriculas, and seem to accomplish their metamorphoses in regions uninhabitable to us. The leaf-wasp (*Tenthredo spinacula*) appears to deposit its larvæ, at a height of nearly 10,000 feet, in the galls of the Alpine rose (*rhododendron ferrugineum* and *rhododendron hirsutum.*)

The coleoptera have also numerous representatives in the region of perpetual snows, with this difference—equally characteristic of other animals—that, upon the southern declivity, they ascend 1000 to 1500 feet higher than on the northern. We may mention, as specially distributed in the topmost zone of the Alpine world :—

The *Chrysomela salicina*, a pretty little beetle, sometimes blue, sometimes deep green, and finely punctuated, which lives almost exclusively upon a species of dwarf willow (*Salix relusa*).

The *Nebria Escheri*, a black beetle, about two thirds of an inch long, with feet and antennæ of a brownish red; and

The *Nebria Chevrierii*, with rust-coloured feet and antennæ, common in the sources of the Rhine.

Special mention must be made of the *Snow-Flea*. Do not think we are referring to an insect of the same species as our common fleas : the snow-flea approximates much more closely to the lice family than to the fleas, though it hops like the latter. The history of its discovery dates back as far as 1839. At this epoch, M. Desor, a learned Swiss naturalist, had undertaken some researches upon the glaciers. Accompanied by some friends, he set out from the hospice of the Grimsel, and arrived in the vicinity of the glacier of the Lower Aar.

He had commenced his observations, when suddenly he heard Agassiz calling him, and shouting, "Come, come, make haste; here are your Mont Rosa fleas." Desor ran to the spot, and saw under a stone the little creatures whom Agassiz persisted in taking to be veritable lice, pretending they had been accidentally brought to these heights.

"I recognised with extreme joy," says M. Desor, "the little creatures whose loss I had regretted a year before. They are not pretty, but, on the contrary, very ugly. However, they showed, in opposition to the opinion of Agassiz, that they really inhabited the glacier, and were not merely chance visitors. We found them by thousands under other stones. . . . Our guide, with whom the glaciers were old acquaintances, had never seen them before, and the tiny creatures excited his astonishment. What surprised us most was the rapidity with which they penetrated into even the most compact ice, till they resembled blood-corpuscles circulating in their vessels. This fact shows that there exist, in the hardest and most transparent ice, certain capillary fissures which escape an unskilled eye: it also proves that the glaciers, on their surface, and down to a certain depth, are by no means incompatible with the development of organised beings."*

The tiny insect in question was at first baptized by the name of *Desoria saltans* (order of the *Thysanouræ* of Latreille), but has since received definitively the name of *Desoria glacialis*. It belongs to the family of the *Poduræ*, singular creatures

* C. Vogt, Agassiz, und seiner Freunde geologische Alpenreisen, p. 181. Frankfort, 1847.

which, by virtue of their form, are a link between the earwigs and the spiders.

These are its generic characters:—

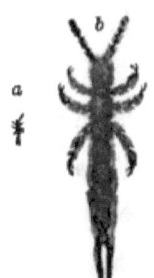

FIG. 15.—*Desoria glacialis*; *a*, natural size; *b*, enlarged.

The body elongated, cylindrical, garnished with long setiform hairs, and composed of eight segments, six of which are perfectly distinct, and two (the two latter) very short, and scarcely perceptible; four-jointed antennæ, longer than the head; long, slender, cylindrical feet; forked tail, silky, and transversely wrinkled; seven eyes, laterally grouped at the base of each antenna; body without scales.

The *Desoria glacialis*, a species at present unique, is of a velvety black, and about one-sixth of an inch in length.

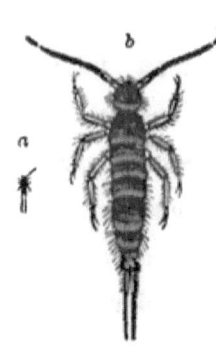

FIG. 16.—*Podura plumbea*; *a*, natural size; *b*, enlarged.

The *Podura plumbea* (or "Spring Tail"), common enough in England, and found under all kinds of stones, will give the reader an idea of the flea of the glaciers.

On comparing these two species, we remark, first, that the *Podura plumbea* is somewhat longer and thicker in body than the flea of the glaciers (see Fig. 16; *a*, natural size; *b*, enlarged); but it is more particularly by the length of its antennæ that we distinguish it. It owes its specific name of *Plumbea* to the livid blue or leaden colour of the scales which cover its body. These scales resemble those of butterflies; only they are much smaller, more finely situated, and very variable in form and size (Fig. 17). In catching it great care is required, for it is so easily crushed; it is, besides, very soft to the touch, though, when examined with a microscope, it is seen to bristle all over with hairs, apparently very hard.

FIG. 17.

Our *poduræ* have also the faculty of leaping, and cling by thousands to humid places, especially to mosses and the under-surface of stones. The mechanism of their leap is explained by the presence of a forked, flexible, and elastic appendage, lodged in a kind of ventral groove beneath the last segments; by projecting this rapidly behind, the whole body of the animal is thrown forward. At the slightest contact the insect folds up its caudal appendage under its belly, and you would then suppose it did not possess one. This circumstance explains why, in many books of natural history of good repute, the *poduræ*, and especially so common a species as the *Podura plumbea*, are represented without this characteristic instrument.

HERBACEOUS PLANTS WHICH BEST ENDURE THE COLD OF WINTER.

The "way to look at things," which is the true foundation of science, varies, not only according to a man's degree of intellectual cultivation, but according to his social condition or profession. The herborist has eyes only for the plants in which he deals,—the "simples" which, as we read in old Gerarde, wrought such wonderful cures in the days of our forefathers,—and from the most exquisite flowers he turns with indifference. The gardener, on the other hand, is wholly absorbed by his love and his hate,—his charming exotics, and his troublesome weeds. The latter he regards with much the same feelings as a society wholly composed of honest men would regard an infusion of the "dangerous elements;" for weeds, like rogues, take what is not their own, and deprive others of their means of sustenance. But to classify plants according to their virtues or vices is not worthy of science, exclaims the rigid botanist. Would you mingle vile self-interest with the pure study of the vegetable kingdom? Remember that all selfish feelings ought to be banished from the sublime sanctuary of analysis and synthesis.

This sounds exceedingly well. Disinterested words, from whatever quarter they come, always produce—perhaps, on account of their comparative rarity—an admirable effect. But what is their real value? To ascertain it, the listener must be able to seize, like so many luminous threads, all the emotions which are acting upon the heart and tongue of the speaker. But we are very far from having arrived at this degree of perfection. Shall we ever attain to it? Yes, because we can conceive its possibility. But, until that golden epoch, the pure love of science will always remain a myth, and we shall not have universally understood the necessity of seeking in the profound study of nature the grand destiny of man.

It is among the weeds and noxious plants that we shall find the species capable of enduring longest the cold of winter. What part, then, do they fulfil in the economy of creation? An ambitious, but not a novel question, which has often been propounded in reference to our parasitical insects.

The best answer which we can make to it is this: Everything invites us to work. Labour is imposed even upon him who least desires it. Earth will yield a return only in proportion to the care we bestow upon her.

If, after having toiled and sown, we had nothing to do but to gather in the harvest, every person would become an agriculturist. But a soil which is not manured will soon grow exhausted; and if it be neither ploughed nor harrowed, instead of barley or vegetables, it will soon be covered with tares; rank weeds will flourish in every field. Such is the chastisement reserved for sloth,—the true "original sin" of the human race.

Well, then, it is among the weeds, everywhere so common, that we meet with the plants best able to brave the rigours of frost.

THE DOG MERCURY.

The annual *Dog Mercury (Mercurialis annua)* is one of the most tenacious. It attracts the passer-by, if he condescend to bestow a glance upon it, only by its extreme abundance; it propagates very largely, though it is by no means partial to all localities. For instance, it avoids the woods as persistently as its congener, the common Dog Mercury (*Mercurialis perennis*) seeks them. It prefers the vicinity of human habitations and uncultivated fields. If let alone, it spreads with a dangerous rapidity, and invades every garden which is not kept in the most exquisite order. Still, we must not deal too harshly with it. It is not altogether unfriendly to man. In truth, owing to its laxative properties, it renders him invaluable services. The country people have great faith in fomentations of Dog Mercury and honey. Understand, we pray you, that not an atom of mercury enters into it, despite its significant name; but a decoction of the annual Dog Mercury, mixed with a little thick honey, answers all the purposes of those lenitive clysters which are so beneficial to excitable temperaments. The leaves of the plant are eaten in Germany like spinach.

Of the *Mercurialis perennis* Mr Sowerby writes :—" This plant was formerly used in medicine, but has long been abandoned as a remedy. It is extremely acrid, and even poisonous, though recommended in some old books as a good pot-herb, probably from being confounded with the annual species. When steeped in water, the leaves give out a fine blue colour resembling indigo. This colouring matter is turned red by acids, and destroyed by alkalis, but is otherwise permanent; it might possibly prove valuable as a dye, if any means could be discovered of fixing it, and the herb has been introduced into this work with the view of drawing the attention of chemists to the subject; no experiments seem to have been lately made upon it."

Let us now advise you how to distinguish our medicinal plant from the "ill weeds" with which it loves to associate.

Its ovate, rough, irregularly-dentated, and petiolated leaves would not give it a sufficiently marked character, had it no other features peculiar to itself. But observe the yellowish-green *glomerules*, arranged, like millet, on a long frail spike. (Fig. 18, *a*.) They exhale, as your nose will inform you, a peculiar aroma, like that of spiced bread: no other plant but our Dog Mercury is gifted with this odour. Now, bring your magnifying-glass to bear upon it; with the point of a knife or a feather open one of the grains which form the *glomerules* of the spike; out of it will leap, as if impelled by an invisible spring, a large number of stamens, easily distinguished by their elastic thread-like anthers, covered with tiny yellow beads. Each greenish grain is a *flower;* the calyx, which also serves as the corolla, is represented by three little petals, forming the external envelope of the little flower. (Fig. 18, *b*.) But something essential is still wanting; in the centre of the stamens you do not find any pistil. Why is

FIG. 18.

so important an organ wanting? Because our little rounded flowers, with their spice-bread odour, have but one sex, are *unisexual;* they are male flowers, since they are furnished only with stamens. In vain do you hunt on the same stem for their companions, the female flowers. You will find them only upon *other* stems, distinct from those which bear the male flowers. The Dog Mercury, then, is a plant whose two sexes are lodged in two different houses, οἶκοι—is, in fact, a *dioecious* species.

But you are sure to find the female flowers in the immediate neighbourhood of the stems with the male flowers. They are easily recognised by their larger and darker leaves (Fig. 18, *c*); and especially by their little twin pods, green, wrinkled, and pedicellate,* which garnish the axil of the leaves. (Fig. 18, *d*.) From this characteristic the female mercury was formerly mistaken for the *male;* and many centuries elapsed before naturalists recognised, what now-a-days seems so simple, that the little pods, joined in couples, and containing each a grain, composed the *fruit* of a single plant; that every fruit proceeds from an ovary; and that every ovary is a sign of the feminine sex.

In the *Historia Naturalis* of Pliny, who was at once so acute and so credulous an observer, we first meet with the name of *Mercurialis*.

"'The plant is so denominated,' he says,† "because it was discovered by Mercury. Its juice, mingled with that of the *hibiscus* (a species of the *Malvaceæ*) and the purslain, forms a kind of unguent, with which, if you thoroughly rub the

---

* A flower with a stalk is called *pedunculate* or *pedicellate*; without a stalk, it is *sessile*.

† Pliny, "Historia Naturalis," xxv. 18.

hands, they can touch molten lead without being injured."

The description which Dioscorides* gives of the *Linozosis*, which he also calls *Parthenion*, or *Mercury's Plant* ('Εϱμοῦ βοτάνον), applies, in the main, to our Dog Mercury. It is true that its leaves "are not like those of the basilic" (φίλλα ὅμοια ὀκίμῳ); but they resemble in all respects those of the smooth variety of cultivated mint; and, apparently, the basilic of Dioscorides was one of our mints. The fruit of the female, he adds,—evidently meaning the male flowers,—are disposed in clusters.

Both species of the Herb or Dog Mercury belong to the family *Euphorbiaceæ*.

Our attention must now be directed to another point. It is a fact, that in winters of moderate severity the Mercury continues to infest our gardens and cultivated fields. It only succumbs to a frost equal to six to ten degrees below freezing-point; then its congealed stem totters, and grows black, and its leaves mingle so completely with the soil that it is difficult to discover any vestiges.

How singular a contrast! The plants most destructive in our kitchen-gardens are frequently the most useful in medicine. There are no drugs more popular than the weeds which we call Herb Mercury, Garden Nightshade, and Dog's-tooth grass. All belong to families whose properties are strongly marked. As already stated, the Mercury ranks among the

---

* Dioscorides, "Materia Medica," iv. 191.

F

Euphorbiaceæ, remarkable for their acrid and more or less purgative juice. In this family occur the most violent drastics, such as the *Croton tiglium*, whose oil (expressed from the seeds) has long been considered an efficacious medicine. The Garden Nightshade is one of the *Solanaceæ*, and cousin-german of the useful potato; and the Dog's-tooth grass, whose roots compose three-fourths of our possets, is of the same family as our cereals.

The Garden Nightshade.

If you have seen—and who has not?—the flowers of the potato-plant, you will immediately recognise the flowers of the Garden or Black Nightshade. (Fig. 19.) This noxious herb—noxious in some, but useful in other respects, and, therefore, not to be visited with too hasty a condemnation—flowers and fructifies throughout the year. Its fertility is extreme; only the severest winter-frosts can crush out its prolific life.

Fig. 19.

Fig. 20.

The fruits which succeed to the flowers are smaller berries or "apples" than those of the potato. (See Fig. 20.)

In the history of botany, and even in that of philosophy,

the Black Nightshade (*Solanum nigrum*) has a certain interest. Thus, says M. Hoefer, both Cordus and Jean Bauhin, botanists of the sixteenth century, have described the flower of this plant as if its corolla were composed of five distinct petals.

Where were the eyes of those great botanists? The corolla of the nightshade, like that of all the *Solanaceæ*, is plainly and obviously monopetalous,—that is, composed of a single piece; to assure yourself of this, you have but to open it out. (See Fig. 21, *b*.) It was the sharp-pointed, ovate divisions of the limb which imposed on the old observers; a fresh proof that *seeing* and *observing* are two very distinct things. Our vision enters into full exercise from earliest infancy; observation is not acquired until after much labour and many years.

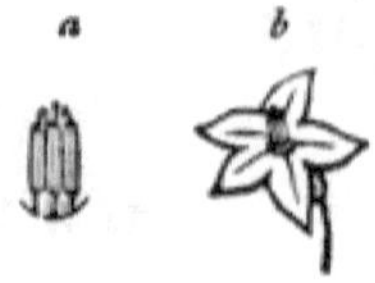

FIG. 21.

Do not forget to add, that the five stamens are brought very closely together by their elongated anthers, as is also seen in the flower of the potato-plant. (Fig. 21, *a*.)

The same botanists who took our solanum for a plant with a polypetalous corolla, considered the Bitter-sweet (*Solanum dulcamara*) to be a metamorphosis of the Garden Nightshade! The former they christened the red-berried solanum (*Solanum baccis rubris*), and the latter, the black-berried solanum (*Solanum baccis nigris*).

But if we once launch into the hypothetical, we shall be unable to stop half-way. If the species of one and the same genus are the result of a transformation, why may we not assert as much of the genera of a family, or the families of an order?

Thus we should arrive, step by step, at an unique type, not only for the vegetable kingdom, but for vegetables and animals, including man himself, and realise, to some extent, the ideal of the Greeks,—*unity in variety.*

Be it acknowledged, however, that *we* have no desire to rise to so lofty an elevation. The potato-plant—unknown to the ancients, inasmuch as it is a native of the New World—has not been found to lose its character since its introduction into the ancient continent; its congener, the nightshade—an old native, like every bad herb—accompanies it everywhere; but its fibrous roots are absolutely virgin of every farinaceous tubercule.

Though the nightshade is common everywhere, Tournefort was the first to describe it with complete accuracy. That great observer even specifies various peculiarities which most of our botanists omit from their descriptions. Thus, he rightly remarks, that the peduncles branch out so as to form a kind of umbel, and do not emerge, as is usually the case, from the axils of the leaves, but a little below, from the very branches of the stem. He was also the first to note—and it was a veritable discovery—that the white flowers of the nightshade, grouped in threes to eights, are each formed of a single cup-shaped leaf,—in other words, that the corolla is monophyllous, and slightly bell-like or campanulated. Nor does he forget to describe the disposition of the five stamens, set close around the pistil, which, as it develops, forms a globular bacciform fruit, embraced by a five-lobed calyx. This fruit, which changes in colour from green to black, is filled with a great number of grains in a thick liquid, exhaling a nauseous odour. As for the leaves, they resemble those of spinach, for which, in some countries, they serve as a substitute.

Like all plants found by the wayside, and among heaps of refuse, the nightshade loves to vary its form, and of its various forms some nomenclators have made as many different species. The typical variety, the *Solanum nigrum*, has glabrous stems and leaves, that is, they are covered with short, but hardly visible hairs; its berries are black.

The smooth variety, or *Solanum villosum*, is rather rare, and has swollen or bulging leaves and stems; its berries are red or of a reddish yellow. The two varieties seem able, by sowing, to be transformed into one another. A sub-variety of the *Solanum villosum* has been described as a peculiar species, under the name of *Solanum miniatum*, so named on account of its vermilion-coloured berries. The *Solanum ochroleucum* and *Solanum luteovirens*, the first with yellowish, and the second with greenish berries, are simply varieties, and the same may be said of the dwarf form, known by the name of *Solanum humile*.

But the physician is more interested in the solanum than either the gardener or botanist. For him it is no useless or noxious weed, but, on the contrary, is an eminently precious herb. And, in fact, if it possessed only one-half the virtues formerly attributed to it, we ought to bow to the ground every time we encounter it.

Listen to our authorities even if you do not respect them.

Cæsalpin asserts that the decoction or juice of the nightshade is a sovereign remedy for complaints of the stomach and the bladder, and regards nightshade-water, mixed with an equal quantity of absinthe-water, as one of the best sudorifics.

Tragus, a physician and botanist like Cæsalpin, recommends the juice of the nightshade as anti-choleraic, as well as efficacious in inflammation of the liver and stomach. And yet, at the same time, he grows emphatic in reference to its poisonous properties. "Do not," he says, "employ this herb immoderately, lest it should happen to you as, in 1541, I saw it happen to an inhabitant of Erbach, near Hohenburg. After having eaten a few nightshade-berries, he was seized, on the following day, by a furious monomania, which led his neighbours to believe him possessed of a devil. After having uselessly employed every kind of exorcism, they sent for me. I made my patient swallow some very strong wine; he fell into a profound slumber; and, when he awoke, was cured." *

Withering affirms that a couple of grains of the dried leaves will act as a powerful sudorific, and that they have also been found useful in some cutaneous disorders.

Here is another authority, before whom naturalists are accustomed to give way. We make use of the *Solanum*

* Tragus, "Historia Stirpium" (ed. 1552).

nigrum, says Tournefort,* when it is necessary to subdue inflammation, or soften and relax the fibres. The pounded herb is applied to hæmorrhoids. The juice, with a sixth-part of rectified spirit-of-wine, is advantageous in cases of erysipelas, ringworm, wildfire, and all diseases of the skin. Nightshade is also employed in anodyne cataplasms.

Tournefort did not confine himself to simple botanical descriptions; he did, what our modern botanists neglect doing,—he made experiments, both physiological and chemical, on the plants employed in medicine. Thus, he began by tasting the different parts of the plant.

"The root," he says, "is almost insipid; the leaves taste like a saltish herb; there is something sharp and vinegary in the fruit; the whole plant has a narcotic odour. The *leaves* do not redden turnsole,† but the ripe *fruit* reddens it greatly; whence we may conjecture that the *sal-ammoniac* contained in this plant is moderated in the leaves by a very considerable portion of fœtid oil and earth, but that the acid portion of the salt is strongly developed in the ripe fruit; so that we must choose our part of the plant according to the purposes we wish to employ it for. The fruits, for instance, are more refreshing, but more repellant, than the leaves, which soften while resolving, cleansing, and absorbing."

---

* Tournefort, "Histoire des Plantes" (ed. 1727), i. 74, 75.

† *Turnsole*, a colouring substance made of coarse linen rags, which, after being cleaned and bleached, are dipped into a mixture of ammoniacal matter, and the juice of the *Crozophora tinctoria*.

We admit that these data leave much to be desired from a chemical point of view. We may well ask, for example, how the illustrious philosopher ascertained the presence of sal-ammoniac in nightshade? But it is not fair to criticise the science of the past, by judging it through the deceitful prism of the science of to-day. We must adopt the methods of our predecessors, when discussing natural productions from all the view-points of their applications.

Dog's-tooth Grass.

In clearing an uncultivated field we uproot a great number of herbaceous plants of different families; but those of the *Gramineæ*, or Grasses, invariably predominate. They are the trailing roots, or *rhizomes*, of certain species which have been included under the general denomination of Dog's-tooth. These tenacious and vigorous roots,—so wholesome in various maladies, so injurious to cultivation,—are, whatever certain botanists may say, far from tracing their origin in all cases to the *Triticum repens* (couch-grass) and *Panicum dactylon*—those terrible enemies of the corn-field, which, once established in the soil, are with difficulty extirpated, and prove very injurious to the "golden crops." Nearly every grass which puts forth *rhizomes* will furnish the *Dog's-tooth*. We may cite, for instance, several species of *Festuca* (as *Festuca rubra* and *Festuca pinnata*), or fescue grass; at least two kinds of meadow grass (*Poa compressa* and *Poa pratensis*), a species of wild-oats (*Avena elatior*), to say nothing of the weeds *Arundo phragmites* and *Arundo epigeios*. The long *rhizomes* of these vivacious plants possess nearly the same properties, due to their saccharine principles.

FIG. 22.—A Corn-field.

How shall we distinguish these plants from one another? Their leaves have almost exactly the same configuration; they are linear;* and their flowers are not apparent,—they do not attract the gaze of the passer-by. Yet they possess all the organs necessary for the reproduction of their species:—three stamens, each composed of an anther and a characteristic filament; on this anther, whose two lobes are arranged like the branches of an X, the pistil softly and tenderly balances itself on the summit of a frail thread, to which it is attached by the back. Remark, too, the two styles with feathery stigmata,† like the barbs of feathers. Nothing is wanting to constitute a complete flower.

There is even a perianth, or calyx, represented by a couple of tiny membraneous scales, scientifically known as *glumellulæ*; then at the base of each spikelet, composed of one or two of these bright green lilliputian flowers, are two other and larger scales, called *glumellæ*: they represent an *involucre*.‡ It is almost unnecessary to add, that the free, unilocular ovary, or seed vessel, forms, as a result of its development, the seed, whose embryo adheres laterally to a farinaceous kernel, or *perisperm*. The union of one or more of these flowers composes a spikelet, and the union of the *spikelets* constitutes the spike, which may be disposed on a simple or ramified axis. Such are, in general, the characters we must keep before us in the difficult study of the Gramineæ.

* Leaves are said to be *linear*, when the veins do not spread out, but run from the base to the extreme point.

† A *stigma* is the continuation of the cellular tissue of the style, and has sometimes projecting cellules of hairs.

‡ A whorl, or ring, of *bracts* (floral leaves) is so called.

Let us now see, more closely, the two plants which, according to the botanists, furnish the root of the Dog's-tooth.

FIG. 23.—A River's Sandy Bank.

When walking along the sandy bank of a river, you must frequently have trodden under foot a low, almost crawling herb, remarkable for its violet-red spikes, which, three to five in number, are arranged like the fingers of a hand, on the summit of a short curving stem.

This glaucous-leaved herb is the *Panicum dactylon* (*i.e.*, fingered-millet) of Linnæus. The long trailing *rhizomes*, joined to some less prominent characters, have been sufficient for some botanists to create a special genus, *Cynodon*, or *Kynodon* (a Greek word, signifying literally "Dog's-tooth"), and to change the Linnean denomination of our grass into

FIG. 24.—(P. 93.)

Cynodon dactylon. It is seldom met with in cultivated land; but in such a locality as we have already described, and sometimes on open sandy shores, where the summer sea comes with a gentle ripple and a subdued music, it may frequently be found. Its long, tough runners creep through and over the loose soil for many yards, rooting at every joint, and furnished with flat, rather short leaves, of a glaucous hue. The flowers grow in narrow, linear spikes, arranged at the top of a short leafy stem in the form of an umbel, and give the grass, when in bloom, a very peculiar and characteristic aspect.

But if the *Cynodon dactylon* is rare in cultivated fields, the *Triticum repens*—commonly called couch-grass, but, in our opinion, the true and genuine dog's-tooth—is particularly abundant. (See Fig. 23.) Its long subterranean stems increase with astonishing rapidity, and if the smallest fragment be left in the soil, it will vegetate, and speedily extend itself, until it becomes almost impossible to extirpate it. It is a kind of wild barley, with stiff leaves of a moderate length, and of a bluish tint, and a double spike, composed of clusters of four to six flowers, each crowned by a narrow ridge. We must not confound the *Triticum repens* with the *Elymus caninus* of Linnæus, which has no trailing underground roots like the former. It differs also from the latter in the roughness of *each* side of its leaves,—only one side of the leaf of the *Triticum repens* being rough,—and in the crests which rise above the flowers.

Was the dog's-tooth known to the ancients? Undoubtedly, for the dog's-tooth flourishes in all climates,—is truly cosmopolitan. But it is difficult to decide whether their *Agrostis* and their *Gramen* apply to the above-mentioned species.

According to Diodorus, the primitive Egyptians lived upon herbs. "They also eat," he says,* "the stems and roots which grow in the marshes. Especially did they

---

* Diodorus, i. 43.

hunt after the *Agrostis*, a plant remarkable for its sweet savour and the sufficient nourishment which it offers to the wants of man. It is likewise considered an excellent provision for cattle, from its fattening properties. It is in remembrance of these benefits that the inhabitants of Egypt, when worshipping their gods, carry this plant in their hand."

The *Agrostis* of Diodorus would apply to all the *Graminaceæ* whose stems and roots contain nutritive and saccharine principles. Let us here remind the reader that the sugar-cane belongs to the same family as barley and the dog's-tooth.

Pliny is much more explicit. What he says of the *Gramen* (or grass), the "commonest of herbs"—*inter herbas vulgatissimum*—and of the geniculated spaces between its knots (*geniculatis serpit internodiis*), applies with tolerable accuracy to our *Triticum repens*. He also speaks of the diuretic properties of a decoction from its trailing roots.* As for his *Gramen aculeatum* (or needle-like grass), it is positively our *Cynodon dactylon*. "The five spurs or needles which shoot out," he says, "from the top of the stem, have procured it the name of *Dactylon*." To these digitiform spikes he attributes the property of checking the bleeding of the nose, when they are introduced into the nostrils. But a thorn is much better fitted to produce this effect; the spikes of the digitated panicle of the *Cynodon dactylon* are much too soft to determine epistaxis by a mechanical action. So it is not improbable that they owe their putative virtue to their colouring,

---

* Pliny, "Historia Naturalis," xxiv. 188.

which is not unlike that of blood, and which has even procured for the species the name of *Digitaria sanguinalis*. In the same manner the capricious mediæval imagination pronounced liver-wort, with its marbled leaves, a sovereign remedy for diseases of the lungs,—organs remarkable for their marbled appearance.

Dioscorides is quite as explicit as Pliny. What the latter names *Gramen*, he, however, calls *Agrostis*. After having particularised the nodosities of the stem—a feature common to nearly all the Graminaceæ—he describes very clearly the long creeping roots put forth by the said stem; and he does not forget to mention the sugary savour, so characteristic of the rhizomes (ῥίζας γλυκείας) of the *Triticum repens*.* Theophrastus confines himself to indicating the *Agrostis* as a herb which infests the fields.†

The *Cynodon dactylon* is, at the present day, very common in Greece, where it is specially partial to low grounds, which are somewhat damp and sandy. The inhabitants call it *Agriada*, a name derived from ἄγριος, "wild." But if we may believe Fraas, the author of a *Flora Classica*, the genuine dog's-tooth, *Triticum repens*, is, on the contrary, very rare in the land of Socrates. This is a curious fact, if a fact, for geographical botany.

Throughout the Middle Ages, and down to the eighteenth century, were confounded, under the generic name of *Gramina*,

---

* Dioscorides, iv. 30.
† Theophrastus, "Historia Plantarum," i. 10; xi. 2, 4.

or grasses, the most diversely-featured herbs, including the dog's-tooth. Tabernæmontanus, Dodonné, Mathiole, Jean and Gaspard Bauhin, were the first to attempt the clearing of a path through this intricate wilderness. They eulogised, at the same time, the emollient properties of the dog's-tooth.

Tournefort[*] and Bernard de Jussieu, who appear to have made a chemical analysis of it, pretend that the roots of the dog's-tooth contain a large quantity of oil, earth, and several acid liquids, as well as a little fixed salt. "According to all appearance," they add, "the roots act by means of a salt analogous to salt of coral, enveloped in a great deal of sulphur."

Instead of mocking us with such fantastic analyses, which can only excite the laughter of our modern chemists, Tournefort and Bernard de Jussieu would have deserved better of science if they had applied themselves to the task of introducing light and order into the cloudy chaos of the *Graminaceæ* of the botanists of their age.

But winter is passing away, and the time for the singing of birds is at hand. Already the earth is awakening from her prolonged lethargy; the hedgerows are green with budding leaves; the purple crocuses shine in many a sheltered field; on bank and brae, in glen and vale, the glory of the primrose makes glad the heart of man; the wood anemone hangs its delicate head in the woodlands; and it seems as if a gladder feeling animated the universal nature.

[*] Tournefort, "Histoire des Plantes," ii. 54.

And the heart and the brain and the soul sympathise in this apparent delight of material things; the heart beating more freely, the brain feeling a stronger working power, and the soul rising to purer views of life and its duties :—

> "Oh, who can speak the joys of spring's young morn,
>   When wood and pasture open on his view,
>   When tender green buds blush upon the thorn,
>   And the first primrose dips its leaves in dew!"

FIG. 25.—On bank and brae, in glen and vale.

BOOK II.

SPRING-TIME.

Now that the Winter's gone, the earth hath lost
Her snow-white robes, and now no more the frost
Candies the grass, or calls an icy cream
Upon the silver lake or crystal stream;
But the warm sun thaws the benumbèd earth,
And makes it tender; gives a sacred birth
To the glad swallow; wakes in hollow tree
The drowsy cuckoo, and the humble bee;
Now do a choir of chirping minstrels bring
In triumph to the world the youthful Spring;
And valleys, hills, and woods, in rich array,
Welcome the coming of the longed-for May.
—Thomas Carew

'Tis silence all,
And pleasing expectation.
  Even mountains, vales,
And forests, seem impatient to demand
The promised sweetness. Man superior walks
Amid the glad creation, musing praise,
And looking lively gratitude.
—Thomson.

CHAPTER I.

WHAT MAY BE SEEN IN THE HEAVENS.

"Blue the sky,
Spreads like an ocean hung on high."
—BYRON.

IT may be doubted whether many of the patrons of Mudie's are acquainted with the works of a philosopher, who, in his day, enjoyed no little fame—I mean, Robert Boyle (1627–1691),—and yet there are passages in them well worth attentive perusal, from the lucidity of their style and the soundness of their reflections. He has, for instance, some observations in his "Considerations on the Usefulness of Experimental Philosophy," which are germane to the general purport and tone of our little book. He remarks, that the contemplation of the vastness, beauty, and regular motions of the heavenly bodies, the excellent structure of animals and plants, besides a multitude of other phenomena of nature, and the subserviency of most of these to man, ought, certainly, to induce him, as a rational creature, to conclude that this vast, beautiful, orderly, and, in a word, many ways admirable, system of things, that we call the world, was framed by an Author supremely powerful, wise, and good.

The works of God, he adds, are so worthy of their Author, that, besides the impresses of His wisdom and goodness that are left, as it were, upon their surfaces, there are a great many more curious and excellent tokens and effects of Divine artifice in the hidden and innermost recesses of them; and these are not to be discovered by the perfunctory looks of oscitant and unskilful beholders; but require, as well as deserve, the most attentive and prying inspection of inquisitive and well-instructed considerers. It is not by a slight survey, but by a diligent and skilful scrutiny of the works of God, that a man must be, by a rational and affective conviction, engaged to acknowledge, with the prophet, that the Author of nature is "wonderful in counsel, and excellent in working."

That He is wonderful in counsel and excellent in working must be the conclusion of every devout student of the celestial phenomena; and to those we shall, therefore, devote the first portion of our Spring meditations.

What reception would formerly have been given to any poet who had dared to exclaim—

> "The bright face of the heavens contemplate,
> And then, as in a mirror, you shall see,
> Outlined, the figure of the rounded earth"?

Would he not have been met with the reproach which Horace, in his *Ars Poetica*, so epigrammatically formulates?—

> "Pictoribus atque poetis
> Quidlibet audendi semper fuit æqua potestas."
>
> An equal licence ever was accorded
> To poet as to painter, that he might
> The boldest sweeps of fancy still essay!

As for men of science, they would not have condescended to honour even with a smile such strange and fantastic words.

Let us suppose, now, that our poetical astronomer, thus contemned, had addressed his scientific censors in some such language as the following:—

Do not think, illustrious sirs, that it is by a purely poetical licence I call the firmament a mirror in which the earth may be seen reflected. Only, to prevent all equivoque, we must understand one another. The mirror to which I am alluding does not reflect *light*, but *movement*. It is in a particular movement of the stars that the true figure of our planet is reflected, is revealed to us. But before the human mind can appreciate this movement,—especially before it can discover the cause,—we must be prepared to devote ourselves to centuries of assiduous effort. In this long interval, philosophers of every class will allow unrestricted scope to their imagination.

Shall we, then, recall some of these opinions,—some of these truly poetical licences?

Homer and Hesiod represented the earth as a disc, or as a flat rondel, surrounded on all sides by a winding river which they called the Ocean, and which, in the extreme East, communicated with the Phasis, in Colchis. Above this terrestrial disc the outspread sky was arched like a vast dome; a dome supported by two massive pillars, resting on the shoulders of the god Atlas.

Surely the ancient poets must have evolved the earth-disc from their own prolific imagination. Can they never have seen a far-off vessel, showing, as it approached them, at first the tops of its masts, then its swelling sails, and finally its hull? They might have made so simple an observation in any seaport; if they did, why did it not suggest to them the idea that the earth, instead of being level, must be round? Because it is easier to let the imagination speak than the reason.

The fiction of the earth-disc remained long unshaken, with the exception of a few modifications. Thales figured to himself the earth as floating on a humid element. And, six centuries later, we find Seneca still adopting the opinion of the Greek philosopher. "This humid element (*humor*)," he says, "which sustains the disc of the earth like a ship, may be, perhaps, the ocean, or a liquid of simpler nature than water."[*]

But how, then, was the rising or setting of the stars explained? The ancients supposed that they were extinguished at sunset, and rekindled at sunrise. Thus, an unfounded hypothesis has for its consequence a still more baseless hypothesis; and in this manner we glide down the slope of fiction to fall eventually into an abyss of contradictions. Such is the true punishment of error.

[*] Seneca, "Quæstiones Naturales," vi. 6.

Let us continue.—According to the Chaldeans, who were thought to be profoundly versed in astronomy, the earth was hollow, and shaped like an egg-shell. And,—adds Diodorus, from whom we have this detail,—they adduce numerous and plausible proofs of this assertion.

Yet was this idea in direct opposition to the evidence of our senses when we travel over a wide plain, or navigate the great deep; at least, unless we admit that the earth has the form of a reversed egg-shell, with its convex face uppermost, and its concave one beneath. Heraclitus of Ephesus introduced the Chaldean doctrine into Greece.

Anaximander represents the earth as a cylinder, whose upper face alone is inhabited. This cylinder, adds the philosopher, is a third of its diameter in height, and floats freely in the midst of the celestial vault, because there is no reason why it should move more to one side than the other. Leucippus, Democritus, Heraclitus, and Anaxagoras,—names of high repute in the history of philosophy,—adopted Anaximander's system, though it was neither more nor less than a wild phantasy.

Anaximenes and Zenophanes, without pronouncing dogmatically on the form of the earth, represented it as resting,—the one upon compressed air, the other upon roots which were prolonged *ad infinitum*. But upon what was the compressed air supported? And of what nature were these mysterious roots?

Plato, with a nearer approximation to the probable, gave to the earth the form of a cube. The cube, bounded by six square equal surfaces, appeared to him the most perfect geometrical solid, and consequently the most suitable for the earth, supposed to be the centre of the universe.

Eudoxes, who, in his long travels in Greece and Egypt, must have seen new constellations rising in the south, while others disappeared in the north, never ventured to adduce from his astronomical observations the sphericity of the earth.

Aristotle, bolder than Eudoxes, was led to the conception of this sphericity by simple consideration of mechanics. The earth, he said, must be a sphere, because each particle of matter is carried, by gravity, towards the centre; and as this fact is general, the superficial particles must be at an equal distance from the centre. This theoretical view was adopted by Archimedes, who applied it to the waters covering the terrestrial surface. Aristotle went further; he saw the rotundity of the earth in the shadow thrown by the latter on the bright face of the moon during its eclipses.

It is a noteworthy fact that the arguments of Aristotle, founded on a method to which all the progress of science is due, remained unaccepted for two thousand years. And why?

We shall attempt to explain.

Among those subjects whose comprehension seems to have been specially difficult to the mind of man, we must include the fact that the earth floats without any solid support in the infinity of space, and carries its denizens on its surface, both above and below.

Our creeds, which have ever pretended to explain everything in the physical as well as in the moral order, have here endeavoured to come to the assistance of the weakness of the human mind. And as each creed asserts itself to be the best, to the exclusion of every other, men began to imagine for the earth a navel, and placed it where it was supposed pleasing to the divinity.

The Greek priests dismissed a couple of eagles, one towards the west, the other towards the east; the place where these favourite birds of Jove, of the "Father of gods and men," encountered each other, was to be considered the "navel" of the earth. It chanced to be Delphos, whose oracle was the most esteemed in the ancient world; the sacerdotal caste accumulated there immense wealth. The Greek priests prudently refrained from dealing with the difficult problem of the earth's solid support.

The Hebrew pontiffs, however, were not so reserved. After having made Jerusalem the "navel of the world," they allowed for the earth itself seven solid columns as a foundation. The question of the Antipodes, in which the greatest intellects of antiquity believed,—Pythagoras, Plato, and Aristotle,—was thus pontifically judged and condemned. And Christians who preferred to follow the Judaical observance of external ceremonies to a true comprehension of the spirit of the Gospels, exaggerated the application of this sentence.

The dogmatic condemnation of the existence of the Antipodes long arrested man in his search for that fourth quarter of the world whose inhabitants should have their feet directed towards ours. It was one of the principal obstacles which Columbus was called upon to surmount in the realisation of his sublime idea. When cited before the Council of Salamanca,—composed of prelates and men of science,—he had to meet the revived objection of Lactantius, a Christian apologist of the third century:*—"Can there be anything more absurd than a belief in the existence of Antipodes, of inhabitants with their feet opposite to our feet, of people who walk with their feet in the air, and their heads on the ground?—that there is a part of the world where everything is inverted, where trees throw out their branches from top to bottom, while it rains, and hails, and snows, from bottom to top?"

Columbus admirably demonstrated, from an artificial globe, that flies walked as easily on the lower as on the upper surface, and hence pointed out that men, compared with the size of the earth, are much smaller still than flies. But his judges persisted in their conviction, and did not fail to cast in his face the jesting words of Plutarch: "Philosophers, rather than renounce a favourite hypothesis, would make human beings crawl on the lower face of the earth like worms or lizards." But it was principally the authority of St Augustine which they invoked to condemn a belief in the Antipodes. St Augustine had declared such a belief incompatible with the dogmas of the faith; for to admit the existence of inhabitable lands, in the opposite hemisphere, would be to

* Lactantius, "De Falsa Sapientia," iii. 24.

admit the existence of peoples not descended from Adam, since it would have been impossible for him to traverse the ocean lying between Asia and the Antipodes !

Some authorities denied the Antipodes on the ground taken by certain classic writers, that the regions of the opposite hemisphere were uninhabitable under the tropics, on account of the extreme heat, and near the Poles, on account of the extreme cold. Others cited Epicurus, affirming that the earth was inhabitable and roofed with a celestial vault only in our western hemisphere, the other half being an inaccessible chaos. Others pretended that no traveller could reach the east by way of the west, because the earth, being pear-shaped, he would encounter on his road an insurmountable tuberosity,—undoubtedly the tail or stalk of the pear! Finally, there were some who dwelt upon the magnitude of the earth, whose circuit would occupy a voyage of upwards of three years.

It was to this objection, as the most serious, that Columbus principally addressed his reply. And he replied by discovering the New World. But had not this daring genius been supported in his projects by the Spanish sovereigns, Ferdinand and Isabella, he would have been handed over to the Inquisition, and condemned as a heretic. It was then so dangerous to believe in the Antipodes, that a Bishop of Salzburg was deposed from his episcopal throne, and deprived of his ecclesiastical dignity, by the Pope Zacharias, for having countenanced the heresy.

We now know why, for a whole series of centuries, men would not follow in the footsteps of Aristotle, who was the first to establish theoretically the sphericity of the earth.

The discovery of the New World, and the voyages of circumnavigation which rapidly succeeded one another, demonstrated, not only that inhabitants there are whose feet are opposite to ours, but that the earth does not rest upon any species of support; that it floats, like a star, freely in space.

The ice was broken. The question of the earth's figure was revived, and, this time, discussed in a new light.

Is the earth perfectly round ?

Copernicus never doubted it; he who was the first, after Pythagoras, to represent our planet as revolving round the sun. The geometrical sphericity of the earth wonderfully harmonised with the perfect circles in which he supposed the planets to move. Kepler, who had first laid a sacrilegious hand on the holy figure of the circle, and on the circular orbits of the stars, never ventured, however, to dispute the perfect rotundity of the earth; it appeared to him a matter beyond all controversy. Galileo was the first to hazard a doubt. But this doubt became a certainty only through the labours of Huygens.

Galileo, who died in the very year that Newton was born (1642), had discovered, as we know, that all bodies, in falling, obey an uniformly accelerative force, called *gravitation*, and that the space traversed increases as the square of the time occupied in their descent. Huygens perceived that gravity varies according to the parallels of latitude, and it was not long before he demonstrated, by the number of oscillations which a pendulum of a certain length performs in a certain time, that it diminishes in a regular ratio as we approach the Equator, where it reaches its *minimum*, and that it augments, on the contrary, in due proportion as we approach the Poles, where it must attain its *maximum*. Strong in this knowledge, and knowing, moreover, that the material molecules, uniformly distributed in the volume of a sphere, act upon a point of its surface as if they were all reunited in the centre of that sphere, Huygens deduced from it the inequality of the equatorial and the polar radius; he attempted even to determine how much the former exceeded the latter. We know, nowadays, that this difference is 139,670 feet (41,848,380—41,708,710 feet).

Newton admits, with Huygens, that the earth bulges out at the Equator and is flattened at the Poles; that, in a word, it is a spheroid of revolution. He went much farther: he made the precession of the Equinoxes depend upon this flattening; but he did not furnish its mathematical demonstration. What has been the result? A free skirmishing ground for all opinions.

While Newton maintained that the form of the earth was that of a spheroid flattened at the poles, as a necessary sequence of the great natural law which bears his name, Jacques Cassini declared himself in favour of an elongated spheroid. The difference between these two illustrious teachers originated a controversy which lasted for upwards of fifty years. The Academy of Sciences of Paris pronounced, not unnaturally, in favour of the opinion of their colleague, though it was far from having the authority of

FIG. 26.—Sir Isaac Newton.

Dominique Cassini, father of Jacques, and, still less, that of the illustrious President of the Royal Society of London. But patriotic ardour supplemented the weakness of their arguments. The flattened spheroid and Newton's law were rejected by France, because they were an English invention. Undoubtedly, no one *openly* acknowledged so paltry a reason, but it was certainly true as a sentiment. As everybody knows, it was Voltaire who first removed the prohibition, and popularised the Newtonian philosophy in France.

How did our astronomers finally succeed in demonstrating mathematically the veritable form of our planet?

To obtain a clear and accurate conception, we are obliged to transport ourselves back two thousand years. Let us recall, in the first place, that, owing to the diurnal movement, all the stars progress from east to west; that they rise and set, to recommence the same rotation. This is a general and conspicuous fact, which everybody can confirm for himself. But now for another, whose observation requires a little more time and patience. During the diurnal movement, which carries on all the stars and the sun himself, the latter progresses independently, in the inverse direction of the celestial vault, as a fly might do upon a revolving globe. But this second fact is complicated with a third: While advancing on his own account, from west to east, the circle which the sun traverses is not parallel to the Equator; the radiant luminary transports himself alternatively into the northern and southern hemispheres, accomplishing this rotation in 365 days and a fraction of a day, in an oblique plane, which cuts that of the Equator under an angle of about $23\frac{1}{2}°$.

Let us here take advantage of a parenthesis to explain a few astronomical technicalities, necessary for the due comprehension of our subject.

It is in the plane, or *oblique circle*,—ὁ κῦκλος λοξός, as Ptolemæus called it,—that eclipses occur, owing to the relative positions of the sun, earth, and moon; and it is for this reason modern astronomers have denominated it the *Ecliptic*. The Ecliptic is the Equator of the *oblique sphere* (σφαῖρα ἐγκεκλιμμένη), properly so called, as the Equator is that of the sphere of the world, or the *right sphere* (σφαῖρα ὀρθή). The circles parallel to the Ecliptic, which continue to diminish in diameter up to the poles of the oblique sphere, bear the name of *parallels of latitude*; and we give that of *meridians of longitude* or *oblique ascensions* (ἀναφοραὶ λοξαί) to the great circles which cut the first rectangularly as they all pass through the axis and the poles of the Ecliptic. The same division by circles cutting each other rectangularly has been made on the right sphere, or sphere of the world. Only, *there* the latitudes are named *declinations*, and the longitudes *right ascensions*. The general diurnal movement is a movement in right ascension; it is measured upon the Equator. The individual annual movement of the sun is a movement in longitude; it is measured upon the Ecliptic.

The zone, or belt, which the sun seems to trace in its annual march, from the limit of its southern excursion (*the winter solstice*) to the limit of its boreal excursion (*the summer solstice*), and in returning from that limit to the other, after having twice passed through the equinoctial line (or Equator), —this zone is marked on the firmament by a belt of constellations known as the *Zodiac*.

These constellations are named, according to the figurative

grouping of the stars (on which we have commented in Book I.),—the Ram, the Bull, the Twins, the Crab, the Lion, the Virgin, the Balance, the Scorpion, the Archer, the Cow, the Water-bearer, the Fishes. There are twelve, three for each season. The constellations represented by these figures, so singularly chosen, spread over the whole celestial vault,—that is, over an extent of 360°.

To resume.

The heavens, like earth, have their annals: everything changes there as in the human world. In the age of Hipparchus,—or some two thousand years ago,—the sun entered, at the spring equinox, into the zodiacal sign of Aries; in the summer solstice, it entered into that of Cancer; at the autumnal equinox, into Libra; and at the winter solstice, into Capricorn. These signs then corresponded exactly to the constellations which they represent.

Now, whatever Aristotle and his disciples may say, the firmament is not incorruptible (ἀφθαρτός) and immovable; even the *fixed* stars, as we call them, change their place in time. We have seen that the whole celestial vault or "right sphere" runs, from east to west, around the poles of the world; we have seen also that the sun moves, on his own account, from west to east, around the poles of the oblique sphere, or the Ecliptic. Well, this does not suffice; there is a third movement to be observed,—that of the right sphere itself round the poles of the Ecliptic; and this, not like that of the sun, from west to east, but inversely, from east to west.

Only, this movement of the starry sphere in longitude, or parallel to the plane of the Ecliptic, is extremely slow, compared with the movement of the same sphere in right ascension, or parallel to the Equator of the world. While the former traverses in twenty-four hours the 360° of the circle, the latter occupies (in round numbers) 25,000 years.

Who was the first discoverer of the slow movement of the heavens? Hipparchus. This great astronomer, on comparing his own observations with the more ancient ones of Aristillus and Timocharis, succeeded in ascertaining that the constellation which, 150 years before him, corresponded to the spring equinox, did not, in *his* time, any longer coincide exactly with the same equinoctial point, but had outstripped or preceded it about 2°. This is what we mean by the *precession of the equinoxes.*

Hipparchus was at first of opinion that this movement affected only the constellations of the Zodiac; but he soon became assured of its universality. He perceived that if it does not alter the parallels of latitude, because it has occurred parallel with the Ecliptic, it makes the position of the equinox retrograde from east to west, and the sun pass slowly through the same constellations in the reverse of the order in which he annually traverses them.

We know this movement now to nearly the fraction of a second. By an inappreciable *daily* quantity, it rises, at the end of the year, to 50″.3,—in a century, to about $1\frac{1}{2}$°,—in twenty centuries, to 30°, or the twelfth part of the Zodiac. It is for this reason the Ram, which, in the days of Hipparchus, was occupied by the sun in spring, has no longer any value as a commemorative sign; it gives place nowadays to the constellation of the Fishes, and corresponds to the constellation of Taurus, or the Bull; the constellation of Taurus to Gemini, or the Twins; the constellation of the Twins to Cancer, and so on. But little more than a month, then (a month of 2000 years!), of the *great year* (a year of 25,000 years!) has elapsed since the epoch of Hipparchus. It is to astronomy especially that, with a slight variation, we may apply the aphorism of Hippocrates—"*Brevis vita, ars longa*" (Life is short, and art long).

The precession of the equinoxes explains why the pole of our starry vault does not occupy invariably the same point of the firmament, and why the constellations which we now see shining during the nights of a given season change their places as time glides by.

But what is the cause of this movement?

Before this question, as before a sovereign tribunal, appear the two opposite doctrines which have been enunciated on the value of the earth and the sun in the world's system. According to the doctrine at once the oldest and most intolerant, the earth occupies immovably the centre of the world; the sun and the planets are only its satellites; they, like the moon, revolve around the earth; finally, all the starry sphere, the whole celestial vault, rotates upon its own axis in four-and-twenty hours. We have been speaking as if this were really and truly the condition of things. If we admit this doctrine, which bears the name of the Ptolemean system,—though, in truth, it is probably as old as humanity itself,—how shall we explain the precession of the equinoxes? We cannot do otherwise than suppose, that while the celestial sphere executes its diurnal movement round the poles of the world, it executes another and much slower movement round the poles of the Ecliptic.

But this assuredly is a most singular supposition. What! the same starry sphere revolves at one and the same time parallel to the plane of the Equator, and parallel to another plane (the Ecliptic) inclined upon the first? After having imagined eight spheres of crystal to explain the movements of the moon, the sun, the planets (Mercury, Venus, Mars, Jupiter, Saturn), and the stars, do we require a ninth? Where will you stop, if you begin to discover additional movements? You are condemned to wander from hypothesis to hypothesis, until you fall into an abyss of contradictions!

Such is the language employed by the tribunal of posterity, in addressing itself to the error which would substitute appearance for reality.

According to the other theory, it is the *sun* which occupies the centre of the system, and it is the *earth* which, accompanied by the remainder of the planets, revolves around *it*. This theory is likewise of considerable antiquity, though generally known as the *Copernican system.* But four-and-twenty centuries prior to the epoch of Nicholas Copernicus, it was taught by the "Samian sage," Pythagoras, and his disciples. The system then in acceptance, however, imposed upon them the necessity of silence. Ptolemæus was acquainted with it, but endeavoured to turn it into ridicule. "There are people," he says, "who pretend that heaven is immovable, and that it is earth which revolves on its own axis; evidently these individuals are unaware how supremely absurd is their opinion (πάνυ γελοιότατον)." And it was in the name of logic and mathematics that Ptolemæus thus treated the Pythagoreans!

In the system of Copernicus,—the diurnal movement of the right sphere,—it is the earth's rotation upon its own axis which, being prolonged into the heavens, marks there, by its extremities, what are called the *Poles of the world,* just as the *Equator of the world* is simply the prolongation of the terrestrial Equator. As for the Equator of the oblique sphere (the Ecliptic), in which the sun *apparently* moves, it is, in reality, the identical plane in which the earth moves during its annual revolution round the sun. Now, in this movement of translation, the axis of the earth does not remain constantly parallel to itself; it deviates,—very slightly, it is true,—and so as to be scarcely perceptible to several generations of men. It is then quite natural that our successors should see, for a long time to come, the northern pole of the starry sphere near the extremity of the tail of Ursa Minor. But, two thousand years hence, this slow deviation will have become very perceptible; astronomers will then see the pole of the world in another constellation, and, as this displacement is continuous, the prolonged axis of the earth will have traced on the firmament, in 25,000 to 26,000 years, a circle parallel to the plane of the Ecliptic, and having for its centre the pole of that plane. This circle is the base of a cone whose summit rests upon the earth. (Fig. 27, *a.*)

But this imaginary defined circle (which appears elliptic on account of the perspective) is but the mean of a series of oscillations around the pole of the world, which changes its position, as we have just shown. (Fig. 27, *b.*) These oscillations originate in the circumstance that the axis of the earth inclines alternately forward and backward, in such wise, that a star, after having approached the Pole, immediately afterwards recedes from it; they cause the terrestrial globe to resemble the head of a man who, by an alternation of gesture, says alternately yes and no. Only, while man (the puppet!) occupies but a second or two in affirming and denying the same thing, the earth employs about eighteen years and a half in inclining once forward to say *yes* (in Latin, *adnuere*), and once backward to say *no* (in Latin,

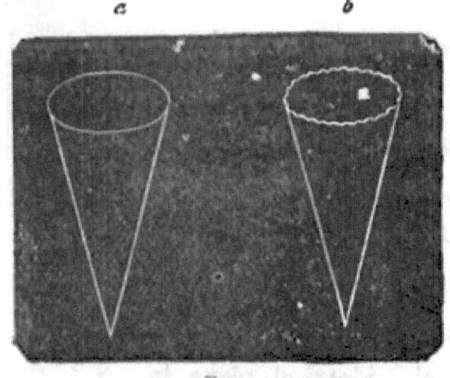

Fig. 27.

abnuere). This is scientifically denominated the *nutation* of the earth.

Who was the fortunate mortal to discover a phenomenon so singular? Bradley, the English astronomer; the same who discovered the *aberration of light*. It was in the course of his researches to determine the annual parallax or distance of the stars that, at an interval of nineteen years, he made, in 1728, the discovery of the aberration of light, and, in 1747, that of nutation.

The reader may not be displeased to know under what circumstances he accomplished the latter discovery. While observing, for several successive years, the circumpolar stars, and notably the star γ in Draco,—a constellation situated between Ursa Major and Ursa Minor (see Fig. 2, p. 9),—Bradley noticed that this star changed its position by a movement constantly directed towards the north, from 1727 to 1736, or for a period of nine years. When it had reached the latter limit, the star appeared stationary for a moment, and then retraced its course in a southerly direction. Would it also occupy a period of nine years to arrive at the limit of this contrary excursion? Bradley affirmed that it would, and communicated his prediction to a French astronomer, Le Monnier.

How was Bradley led to appear in the new character of a seer?

By two special circumstances—the universality, and the duration of the phenomenon.

If the star γ in Draco had been the only one to direct its course towards the north, Bradley would probably have been led to believe that the Pole exercised upon it a peculiar attraction; but he perceived that many other stars rose in like manner towards the Pole with an uniform and constant march; it was, therefore, more natural to suppose that the Pole advanced towards *them*. And what strengthened the probability of this hypothesis was, that the stars situated in the neighbourhood of the *course* of the solstices exhibited a corresponding displacement. But there was already recognised as in existence a peculiar movement which explained the precession of the equinoxes. Was it necessary, therefore, to suppose a second, a kind of rotatory movement? Newton had already thought of it, by imagining a nutation, through which the Pole might alternately rise and sink on the plane of the Ecliptic in the space of a year. But the displacement which occurs in that interval is too slight to be perceptible to observation. There might, therefore, be a reasonable doubt of the accuracy of Newton's idea.

Bradley resumed the idea of his illustrious compatriot. He recognised in the northward movement of the stars the effect of a similar rotation, but one which took much longer in its accomplishment. By doubling the interval of nine years, to the term of which he had seen the movement become stationary, he obtained a period nearly approaching that which the moon employs in returning to the same nodes. This coincidence flashed upon him like a ray of light.

We must here remind the reader,—who, we hope, is not

weary of our scientific or semi-scientific disquisition,—that the lunar nodes,—*i.e.*, the points of the Ecliptic through which the moon passes when it proceeds from south to north (the ascending node), and from north to south (the descending node),—are the analogues of the solar equinoxes; the equatorial points through which the sun passes on its course from south to north (the spring equinox), and in returning from north to south (the autumn equinox),—points of intersection whose retrogradation constitutes, as we have seen, the precession of the equinoxes. Well, the moon's nodes retrograde in a similar manner by a movement directed from east to west; only it is a much slower movement. While the equinoxes are displaced but fifty seconds (50″) in a year, the lunar nodes, during the same period, and in the same direction, move over a space of 19° 20′ 29″; so that, in less than nineteen years, they have made the complete circuit of the heavens, to return to exactly the same point, after traversing 360°.

Thus, then, we have explained the data on which Bradley rested his prediction. It was confirmed by Le Monnier, who observed, in fact, that the star γ in Draco, and the neighbouring stars, observed by Bradley from 1727 to 1736, moving from south to north, occupied the same period of time, from 1736 to 1745, in accomplishing an equal excursion in a contrary direction, from north to south. These observations enabled him to fix approximatively the quantity of the nutation.

To sum up; it was recognised that the angle made by the axis of the terrestrial poles with the axis of the poles of the Ecliptic, far from remaining constantly equal to itself (the amount was 23° 27′ 30″ in the middle of the present century), varies by 0″.48 yearly, and that this angle itself experiences a variation whose mean value is 48″ in a century. It sometimes exceeds this mean value, and sometimes falls below it, by an amount which rises to nearly 9″.65. Thus, while describing, in an interval of 25,000 to 26,000 years, its curve around the poles of the Ecliptic, the earth's axis describes, from east to west, a small ellipsis in the space of about eighteen and two-third years, and imperceptibly changes, moreover, its angle of inclination.

But, in fine, what is the true cause of all these movements?

Were the earth a perfect sphere, were all its radii of equal length, the effect of the universal *ponderation* would make itself felt as if all the material molecules were concentrated at a single point—the centre; and, apart from this ponderation, which exercises itself in the direct ratio of the masses, and in the inverse ratio of the square of the distances, nothing exists which would sway our globe in one direction rather than in another,—no precession of the equinoxes would take place, the plane of the Ecliptic would invariably coincide with the plane of the Equator, and an eternal spring would smile on the fortunate earth. The dream of the poet would be realised, and light would spread

"Through all the seasons of the golden year."

But, as observation shows, the contrary has taken place, since, besides its movements of diurnal rotation and annual revolution, the earth has its mobile axis, which is independently inclined and displaced. Thus, the material molecules of the planetary surface are not all at an equal distance from the centre; and, consequently, the earth is not a perfect sphere. It is, as D'Alembert has demonstrated, the bulging, equatorial portion which experiences, owing to the solar attraction, a retrograde movement, carrying onward the rest of the globe in a general march, called the precession of the equinoxes.

But this general movement, as we have seen, is, in itself, simply the mean of a series of oscillations, which D'Alembert has also connected with gravitation. He has shown that the nutation of the earth's axis results from the moon's attraction on the bulging portion of our globe. Finally, it has been mathematically demonstrated that the said bulging portion of the earth produces, under the continuous action of the sun, the precession of the equinoxes; just as this portion determines, by its continuous action, the nutation of the lunar axis. As in this universal ponderation all the wheelwork of the world catches (*tous les rouages du monde s'engrènent*), and the planets, such as Mars and Venus, must also have their share in the action, however weak it may be, we have contrived to render an exact account of the slow changes of the obliquity of the Ecliptic.

Let us resume. Movement and matter, all is *ponderated*.

Inasmuch as matter is unequally distributed around the

earth's centre, being flattened at the Poles and bulging at the Equator, it follows that the sun's enormous weight makes it vacillate, so that it describes at its axis a cone around the poles of the plane of its orbit. Its movement we see in the heavens in the precession of the equinoxes. But the terrestrial axis traces it tremblingly, because the moon, owing to its vicinity, exercises a perturbing action on our planet, which, in its turn, exercises on the moon a still more energetic influence.

CHAPTER II.

WHAT MAY BE SEEN UPON THE EARTH.

> " There lives and works
> A soul in all things, and that soul is God."
> —COWPER.

WE have returned, at least in an astronomical sense, to the budding, happy, radiant spring; the sun, in its apparent course, crosses the equinoctial line; the duration of the day, transiently equal to that of the night, will augment in proportion as the great luminary describes above our horizon greater and yet greater arcs of a circle. Yet this is not the budding, happy, radiant spring of the poets. No, if it be spring according to the law of universal gravitation, it is winter still by the law of life. The forest trees, such as the oak, the ash, the fir, and the beech, continue to present the image of death; and the sap which should reanimate them has not awakened from its winter sleep.

A solemn moment is it when the sap—that life-blood of the plant—arrested by the icy grasp of winter in its circulatory movement — receives a new impulse through the

Fig. 28.—Landscape in Winter.

vivifying action of the central luminary of our system. What a subject for study and reflection!

It has been very finely dealt with by Longfellow.

How wonderful! he exclaims,—and we only regard the wonder with indifference, because it is repeated annually,— how wonderful is the advent of spring!—the great annual miracle, as he calls it, of the blossoming of Aaron's rod, repeated on myriads and myriads of branches!—the gentle progression and growth of herbs, flowers, trees,—gentle, and yet irrepressible,—which no force can stay, no violence restrain,—like the influence of love, which wins its way, and cannot be withstood by any human power, because itself is divine power. True enough it is, that if spring came but once in a century, or burst forth with the terror of an earthquake, and not in silence, what wonder and expectation there would be in all hearts to behold the miraculous change! But now the silent succession suggests nothing but necessity. To most men, only the *cessation* of the miracle would be miraculous, and the perpetual exercise of God's power seems less wonderful than its withdrawal would be.

May we venture on another quotation? We take it, gentle reader, from a living poet, whose works are not so widely read as their genuine poetical feeling and wealth of language deserve—I mean Sydney Dobell.

After describing the return of Spring, and her grief and astonishment at the spectacle of earth, pale, frozen, seemingly dead, he continues,—

"She fell upon
The corse, and warmed it. The natural earth,
Which was not *dead* but *slept*, unclosed her eyes;
Then Spring, o'erawed at her own miracle,
Fell on her knees.
   Meanwhile the attendant birds,—her haste outstript,—
Chasing her voice, crowd round, and fill the air
With jocund loyalty.
            With flowers Spring dressed the Earth;
Then did her mother, Earth, rejoice in her;

Fig. 29.—" The attendant birds crowd round, and fill the air."

> And she, with filial love and joy, admired,
> Weeping and trembling, in the wont of maids.
> Meantime her pious fame had filled the skies.
> He that begat her, the almighty Sun,
> Passing in regal state, did call her ' child,'
> And blessed her and her mother where they sat—
> Her by the imposition of bright hands,
> The Earth with kisses. Then the Spring would go,
> Abashed with bliss,—decorous in the face
> Of love parental. But the Earth stood up,
> And held her there ; and, these encircling, came
> All kind of happy shapes that wander space,
> Brightening the air. And they two sang like gods
> Under the answering heavens."

We think that the ancients, if they had seriously reflected upon the important part played by the sun in the economy of nature ; how it is the heart, and spring, and inner power of every movement and manifestation of life ; how it is, as Sir David Brewster says, the centre and soul of our world, the lamp that lights it, the fire that heats it, the magnet that guides and controls it, the fountain of colour, which gives its azure to the sky, its verdure to the fields, its rainbow-hues to the gay world of flowers, and the purple light of love to the marble cheek of youth and beauty ;—we think that the ancients, if they had thought upon, or had known, all this, would not have given the earth a chief place in our system. And that they did so is all the more strange when we remember that they attributed to the world a *soul* (the "soul of the world" is a favourite idea with the great philosophers of antiquity), and looked upon the planets as living creatures.

But they were swayed by that egotistical instinct which leads man to refer everything to himself, even the very gods which he has created after his own image. The Bible teaches us that there is but one God. Alas! are there not as many gods as there are men? Does not each of us create a deity in accordance with his own inclinations, his mode of thought, his degree of mental culture, the sphere of his ideas? Is the God of a tolerant philosopher identical with that of a bigoted fanatic? It is not so much due to a deceitful appearance, an optical illusion, as to a kind of innate infatuation, that the human race have come to consider the planet they inhabit as the centre of the universe.

Causes of the Circulation of the Sap.

Let us return to the sap, the life-blood of vegetation.

How is it that its movement does not recommence at the same time in all plants? Why are some clothed with leaves when the others are scarcely budding? Wherefore, in certain genera, do the flowers appear before the leaves?

Some authorities assert,—but facts show it to be a purely gratuitous supposition,—that the flower, which, with the fruit, seems to be the goal or object of vegetation, demands a greater activity on the part of the sap. But, in truth, many trees and shrubs, such as the poplar, the willow, and the hazel, flourish at an epoch when the sap is barely aroused from its protracted lethargy.

These are questions which have still to be answered.

But upon yet another question we may dwell at some detail. What is the cause of the circulation of the sap?

To the best of our knowledge, this important problem has never been propounded as it should have been. And for this reason: all observers who have taken up its consideration have had in view only the *rising sap*, and the cause of its rising. Evidently this is but a *part* of the problem. The ascending sap, after undergoing an important modification in the leaves, becomes the *descending* sap; just as the venous blood is transformed, on coming into contact with the air in the lungs, into arterial blood. It is this alternative movement of *going and coming* which constitutes the circulation both of the sap and the blood, and which ceases completely only with the life of the plant or the animal. We must, therefore, bear in mind,—which has not been hitherto done,—these two opposite, yet indissolubly connected, movements, before we can approach with advantage the solution of the proposed question.*

Science consists in discovering, among the different ways of looking at things which present themselves to the mind, the one which appears to explain most clearly the phenomena submitted to observation. He who doubts the accuracy of our remark need only join us in reviewing the different opinions enunciated up to the present time on the cause of the rise of the sap.

---

* The best means of ascertaining the coexistence of an ascending and descending sap have been indicated in "The Circle of the Year," pp. 163-8.

Grew, an English botanist, a contemporary of Newton, and his fellow-member in the Royal Society of London, attributed the rise of the sap to the play of the utricles of which the plant is composed. These utricles, he said, maintain a close intercommunication; through their contraction, the sap passes from the lower to the upper, and thus arrives almost at the top of the plant. Grew's authority carried conviction to the minds of many botanists, particularly to those of his compatriots. Yet was his opinion altogether imaginary; the supposed contraction of the utricles does not exist.

La Hire, a French botanist, who flourished at the beginning of the eighteenth century,—son of the geometrician of the same name,—pretended that he could account for the rise of the sap by the play of the little valves with which the interior of the sap-vessels was furnished; at the same time he assigned a very active *rôle* to the fibres of the roots. The fibres elevate, he said, the whole column of superimposed liquid, incessantly introducing, by a kind of suction, new fluids into the organs.

Unfortunately, the "play of the little valves, with which the interior of the sap-vessels is furnished," is a pure invention of La Hire's. Instead of growing wise by experiment, he suffered himself to be led astray by a false analogy. Valves are found in the veins of man and the mammals; but no one has ever seen the sides of the vessels of a plant garnished with valves to induce a circulation of the sap.

Mariotte, so well known by his researches into the compressibility of air, represented the rise of the sap as dependent upon what he called "the attraction taking place in the narrow tubes"—upon what, in fact, we now term *capillarity*. "This first entrance of water into the roots is in obedience," he said, "to a law of nature; for wherever very narrow tubes exist which touch the water, it enters into them, and even rises, contrary to its natural inclination."

Many botanists adopted the opinion of Mariotte. But if it were well founded, all *capillary* bodies, even inorganic ones, ought to present a circulatory movement analogous to that of the sap. Now, this is not the case. A body must be animated, must be living, for attraction to take place in the narrow tubes, and to produce a movement comparable to that of the nutritive liquid.

Malpighi attributed the rise of the sap to the alternating rarefaction and condensation of this liquid by heat; Perrault, to a kind of fermentation; De Saussure, to a peculiar irritability of the vessels. Of these three hypotheses, the first is purely physical; the second, chemical; the third, vital. So, as we see, there is something for everybody—*chacun à son gout!*

The same question has, in our own day, been taken up from a new point of view, on the occasion of Dutrochet's discovery of the endosmose. This philosopher was one of the first to perceive that two liquids, separated from one another by a membrane, quickly effect or induce a current which always carries the thinner liquid towards the denser, and ends by mingling the two completely. "It is endosmose," he said, "which produces at one and the same time the progression of the sap by *impulsion*, and its progression by *affluxion*. The sap would receive its impulse in the spongioles of the roots; thence would be carried towards the upper parts by the turgescence of the organs—by the affluxion, which would thus act as a forcible mode of suction."

The basis of this theory is, that the sap contained in the upper parts will be more concentrated or denser than that in the lower portions of the same plant. But this is a mere supposition. And even this supposition has been swept away by the recent experiments of Hartig and others, which show that the difference in density between the two saps is not only almost null, but in many ligneous plants the lower sap is, on the contrary, denser than the upper.*

Finally, and more recently, M. Joseph Boehm has put forth a theory which offers some points of analogy with that of Grew. According to Boehm, the rise of the sap is the effect of a suction, the cause of which must be sought both in the atmospheric pressure and in the transpiration which takes place through the organs, and notably through

* See the *Botanische Zeitung* ("Botanical Gazette") for the years 1853, 1856, 1859, and 1861.

the leaves of the plant. The part which he attributes to the cellules, of which the organs are composed, he thus describes:—"When the superficial cellules of the plant lose water by transpiration," he says, "of two things, one will happen: either these cellules will contract and shrivel, or they carry up, by a kind of aspiration, to the neighbouring cellules, situated in deeper layers, a quantity of water equivalent to what they had lost. In the normal condition, the latter is always the result; each cellule takes from its neighbour what itself has lost, and this action, becoming more and more general, is continued from the leaves to the extremities of the roots. The cellules of the spongioles replace the water which they have yielded, from the humid medium surrounding them."

In support of this theory,* M. Boehm has made several experiments, which, we fear, will not carry conviction to every mind.

In the different theories which we have been attempting to explain, their authors, as it seems to us, have neglected an essential element—the *life* of the plant. Then, the experiments undertaken by way of proof, have been made upon cut stems or branches, which, consequently, did not enjoy their integral vitality. In fact, the results indicated could just as well have been obtained with inert as with living matter.

* Boehm, "Sur la Cause de l'Ascension de la Séve," Mémoire communiqué à l'Académie des Sciences de Vienne, juillet 1863.

Taking into account all these considerations, we are doubtful whether any value can be placed on the theories just enunciated. Undoubtedly, physical causes, such as capillarity, heat, evaporation, atmospheric pressure, electricity, have a certain marked and constant action. But this action is here complex; it is found combined with a new force, whose effects constitute precisely the profound difference which exists between the massive mineral framework of the globe and the transitory beings peopling its surface. It matters little whether we call it *vital force*, or otherwise; sufficient that it *exists*. We must, therefore, allow for its influence when endeavouring to explain the varied movements of which plants, as well as animals, may be the seat.

A.—Plants.

The Daisy (*Bellis perennis*).

> "Wel by reason men it call maie
> The daisie, or els the eye of the daie." *

Among all the treasures of the floral world, that which should excite in each of us the tenderest emotion, and most readily stir up in our minds thoughts too deep for tears, is the Daisy,—that favourite of our innocent and happy childhood.

Ah! would we were now as content with simple joys as in the days when that wee, modest, crimson-tipped flower was to us a beauty, a prize, and a charm!

* Chaucer.

We wonder how many of our poets have done homage to the sweet and simple "nursling," or rather, whether by any

FIG. 30.—"The Daisie scattered in each mead and down."

true poet it has been neglected. Cowper reminds us that in

"The spring and play-time of the year"

the village-wife and her little ones go forth to

"Prank their hair with daisies."

James Montgomery, whose admiration of nature is somewhat frigid, can yet remind us that—

> "The rose has but a summer reign,
> The daisy never dies."

Chaucer warms into enthusiasm when he thinks of its pastoral, innocent gracefulness (" simplex munditiis ") :—

> " So glad am I when in the Daisy's presence,
> That I am fain to do her reverence ;
> For she of all sweet flowers is the flower
> With virtue filled, and honourable power ;
> For ever fair alike, and fresh of hue,
> As well in winter as in summer new."

Let us not omit a reference to quaint but genial William Browne :—

> " The Daisie, scattered on each meade and down,
> A golden tuft within a silver crowne :
> Fair fall that dainty flower ! and may there be
> No shepherd graced that does not honour thee ! "

Yes! let no poet be taken to your heart of hearts who has no love for the "flower white and rede,"—in French, called "La Belle Marguerite,"—

> " The op'ning gowan, wet wi' dew,"

—Burns's "bonnie gem,"—the flower of the meadow and the lea, of the woodland and the vale.

A modest, unassuming flower, destined to be trodden under the feet of the thoughtless, it withstands the rigorous breath of winter, is beautiful throughout the circle of the year —*Bellis perennis*, as the Swedish botanist not infelicitously called it. Its vegetation is arrested only during the harshest frosts; but it resumes its living growth as soon as it becomes sensible of the first rays of the spring-time sun.

It is at the moment of nature's awakening, about "the solemn Easter-tide," that this "sweet nursling of the vernal year" displays all the simple coquettishness of its chaplet of flowers,—that chaplet which has also procured for it the name of the tiny "Marguerite,"—that is, "little pearl,"—a name which the French have adopted from the Latin—*Margarita*.

Here let us pause, and propound a question.

How would you propose to test the real character, the genuine nature, of friend or acquaintance?

Your answers, dear readers (believe me, I hear them clearly!), are very various. Some of you say, that the best means of sounding the depths of the human heart is by bringing before it a misery which needs to be relieved. Others recommend the bestowal of a benefit. But such processes of analysis appear to me far from being infallible; too wide a margin is left for the operation of sentiments of pride or vanity. Why not conduct the man whose real character you wish to discover into a meadow enamelled with sparkling daisies? Thus you would impose upon nature the task of interrogating him. If he manifest feelings of indifference, you will do well to regard him with suspicion: take care how you admit him into your intimacy; for his heart must be cold, and his mind troubled—

> "The man that takes not daisies to his soul
> Is fit for treasons, stratagems, and spoils."

But to return to our daisy. Observe how, by its organs,

it yields itself—in anticipation, as it were—to its fate. And, first, its long and fibrous roots anchor it so solidly in the soil that the cattle which browse it cannot tear it up. Next, its stem is so short that it seems to be blended with the roots; one might almost doubt whether any existed. But, if you look at it more closely, you may readily assure yourself that the stem is the point whence issue the recumbent branches which bear the leaves. Why does not the daisy boast of stems erect and free? Would they not be incessantly bent or broken by the merry troops of children who love to play and dance upon Nature's carpet, the soft green sward?

The leaves of our daisy, then, seem to issue directly from the roots, without the intermediary of an *apparent* stem, which must not be confounded—recollect this, dear reader!—with the stalk or peduncle that bears the crown of petals. These leaves in form resemble tiny crenelated spatules, with the handles flattened, and the edges trimmed with little hairs or fibres. The peduncle, too, seems to start immediately from the roots. The principal part of the peduncle is surmounted, as already hinted, by the flower, to which we next direct our inquisitive and searching gaze.

What shall we call it? To what shall we liken it? To a gilded button framed in a pearl. This button, this "yellow eye," as Tabernæmontanus, a botanist of the sixteenth century, named it,—the "eye of day" of our old poets,—a drop of gold in a rim of silver,—is not like any other flower; is quite a world or system of Lilliputian blossoms, each of which is represented by a miniature tube, yellow at the summit, and of a greenish white at the base,—the said tube being the union, or combination, of the tops or summits forming the central gem, the gilded button, the drop of gold. You may readily note this arrangement in the larger variety of daisy, the *Chrysanthemum leucanthum* of Linnæus,—two Latinised Greek words which signify, literally, "golden-blossomed white flower."

If you doubt whether each of these tiny tubes be a flower, you have only to analyse them with the assistance of your ever-useful lens. The analysis of one will suffice; for all the others resemble it. Now, with your penknife, split the tube throughout its entire length: you will thus lay bare all the parts which enter into the composition of a veritable flower, commencing with the most conspicuous. Through the magnifying glass you can see five stamens,—free as regards their short filaments, but united by their elongated anthers; a characteristic which gives name to the great family of the *Synantheraceæ*, of which family our daisy is an honoured member;—a bifid (*i.e.*, cloven in two) style traversing the middle of the anthers, which form for it a kind of sheath (see Fig. 31, *a*);—a monopetalous, tubular, and obscurely bilabiated (two-lipped) corolla, inserted at the summit of an unilocular (one-lobed) ovary, which is attached to the calyx (see Fig. 31, *b*). In these tiny flowers, then, which we call in Latin *flosculi*, in French *fleurons*, in English *florets*, nothing is deficient. As they are shaped like tubes, we call them, by way of distinction, *tubulifloral*.

But what are these white rays, lightly shaded with pink, which enclose or encircle the florets? (See Fig. 32, *a*.) Examine them at their points of insertion. You will perceive there some traces of reproductive organs, among which the style is most prominent. As for the corolla, it is represented only by its brown lip, which is immeasurably developed. It is this exaggerated development which constitutes the *white rays*, or petals, that prove so attractive to the eye. (See Fig. 32, *b*.) Do not forget to observe, by the way, that they are rose-tinted only on the side which directly undergoes the action of the light. To distinguish them from the tubular florets,—the *tubuliflora*,—these "white rays" have been called *ligulate florets*, or *liguliflora*.

The complete flowers (or the florets) and the rays (or partially abortive flowers) form, in their aggregate, what our botanists have agreed to call an *inflorescence of the capitula*. Disposed quincuncially on an ovoid receptacle, or *phoranth*, both are grouped (Fig. 32, *c*) in alternating rows.

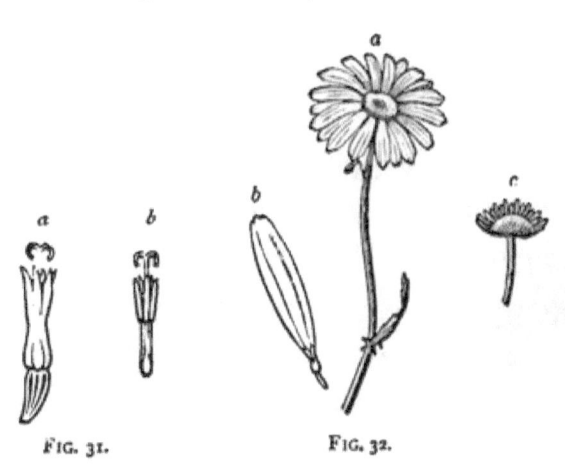

FIG. 31.   FIG. 32.

To explain thoroughly this species of inflorescence, we will venture upon an hypothesis. Let us suppose that we could elongate the said ovoid receptacle as if it were a ball of wax, —it would be changed into a sheath-like inflorescence; all our smaller florets, whose union composes what is improperly called the *flower* of the daisy, would be ranged around an elongated, instead of being placed upon a flattened axis. This axis characterises all the Synantheraceæ of the family of the Compositæ (a sub-order); sometimes naked, sometimes garnished with varied hairs, either shrunken or persistent, it has furnished several characters useful in the classification of genera and species. But possessed with a mania for complicating everything, botanists designate it indifferently *receptacle, phoranthe, clinanthe*, etc. Why not employ one and the same word to distinguish one and the same thing? Why not have preserved the name *axis*, and have attached to it such qualifying terms as might be necessary to indicate simple differences of forms?

The ancients looked upon nature,—I cannot sufficiently insist upon this theme,—with quite other eyes than we do. The study and description of characters, so indispensable to our classifications and nomenclatures, appeared to them a useless labour; they had not even an idea of its value. But it was of signal importance to them to investigate the virtues and properties of plants, so far as they might be rendered available for the preservation of health and the cure of disease.

Our daisy is common in Greece. Theophrastus, therefore, ought to have known it, though he does not refer to it. It is common also in the plains of Italy. Pliny was the first to describe it, under the name of *bellis;* he attributes to it the properties of the St John's wort.* And it is noteworthy that the daisy belongs to the same family as the latter; a circumstance certainly not known in the days of Pliny.

The botanists of the sixteenth and seventeenth centuries are by no means niggards in the eulogiums which they lavish on the medicinal properties of our graceful Synantheracea. Bock (better known, perhaps, under the name of *Tragus,* a Goat), who mistook the yellow anthers for the seeds, recommended the leaves of the *Gänzeblume* (goose-flower, as he called it) as a laxative. Tabernæmontanus prescribed them as a remedy for cramps in the stomach and the spitting of blood.

Ray, who expresses his astonishment that the Greeks had not spoken of it, looked upon the daisy as an excellent vulnerary. "Externally," says he, "we employ it with success in the form either of a poultice or a fomentation; for internal treatment, we mix its juice with vulnerary potions."— These properties procured it the name of *Consolida minor,* which would make it the pendant of the larger Consolida, *Symphytum officinale,* a species of the Boraginaceæ, very common in damp and shady localities.

Ruel recommended cataplasms of daisies and cowslips for gout and scrofulous tumours. Chomel affirmed that he knew by experience that the flowers of the daisy and the *herb robert* † (*Geranium Robertianum*), if dried in a hot

---

* Pliny, "Historia Naturalis," xxvi. 5.
† See "The Circle of the Year."

dish, and applied to the head, considerably relieved headache.*

Wepfer set great value on a mixture of daisy, cress, and rummularia in the treatment of pneumonia; and Michaelis assures us that he had cured dropsies by the use of the flowers of the daisy cooked as a broth.

Tournefort, who was very partial to this kind of observations,—now repudiated by our botanists,—says, that the daisy, taken as a warm drink or a decoction, quickens the blood when congealed by a very severe attack of cold, as happens in pneumonia; it removes obstructions, facilitates the circulation, and gives the fibres an opportunity of recovering their elasticity.†

Garidel sums up in the following words the result of his personal observations:—"I have frequently remarked that the juice of the daisy acts as a laxative, and even as a purgative; the decoction does not have that effect so often as Schroeder observes, who says that mothers frequently give the leaves as a gentle aperient to their children. . . . Care should be taken not to administer this remedy indifferently to all pleuretics, nor at any season; for if we give it when the expectoration is easy, we run the risk, by the employment of a laxative at a wrong time, of spoiling everything, and checking the expectoration. This I have seen occur in several cases, where the remedy had been administered by a hermit." ‡

---

* Chomel, "Histoire des Plantes Usuelles," ii. 282.
† Tournefort, "Histoire des Plantes," i. 103.
‡ Garidel, " Histoire des plantes qui naissent aux environs d'Aix," p. 56.

Can it be true that the commonest plants are the most useful? Nature is quite capable of affording us these surprises; nature, who, by her shifting and proteiform movements, never ceases to laugh at human theories. But men, as was said long ago, have eyes, though not to see; and everybody also knows, from his own experience, that he has ears, not to understand!

However this may be, the daisy, which, as we have seen, was formerly so extolled for its officinal properties, is now-a-days completely ignored by physicians. What, then, are we to conclude? That all the remedies in vogue—melancholy to confess!—are an affair of fashion. When men shall have resumed perukes, and women abandoned chignons for furbelows, we shall remember, perhaps, the virtues of the lowly, tender daisy.

We cannot take leave of our favourite wild-flower without repeating Wordsworth's beautiful stanzas. He takes as his motto a fine passage from Wither, quaint old George Wither:—

> "Her [the Muse's] divine skill taught me this,
> That from everything I saw
> I could some instruction draw,
> And raise pleasure to the height
> Through the meanest object's sight.
> By the murmur of a spring
> Or the least bough's rustelling;
> By a daisy whose leaves spread
> Shut when Titan goes to bed; . . .
> She could more infuse in me
> Than all Nature's beauties can
> In some other wiser man."

On this hint our great meditative poet speaks, and speaks most tenderly and truly :—

> "In youth from rock to rock I went,
> From hill to hill, in discontent
> Of pleasure high and turbulent,
>   Most pleased when most uneasy ;
> But now my own delights I make,—
> My thirst at every rill can slake,
> And gladly Nature's love partake
>   Of thee, sweet daisy ! . . .

> "By violets in their secret mews
> The flowers the wanton zephyrs choose ;
> Proud be the rose, with rains and dews
>   Her head impearling ;
> Thou liv'st with less ambitious aim,
> Yet hast not gone without thy fame ;
> Thou art, indeed, by many a claim,
>   The poet's darling.

> "If to a rock from rains he fly,
> Or, some bright day of April sky,
> Imprison'd by hot sunshine lie,
>   Near the green holly,
> And wearily at length should fare ;
> He need but look about, and there
> Thou art ! a friend at hand, to scare
>   His melancholy.

> "A hundred times, by rock or bower,
> Ere thus I have lain couch'd an hour,
> Have I derived from thy sweet power
>   Some apprehension ;
> Some steady love ; some brief delight ;
> Some memory that had taken flight ;
> Some chime of fancy, wrong or right,
>   Or stray invention. . . .

> "Oft do I sit by thee at ease,
> And weave a web of similes,

> Loose types of things through all degrees,
>   Thoughts of thy raising;
> And many a fond and idle name
> I give to thee, for praise or blame,
> As is the humour of the game,
>   While I am gazing.
>
> "A nun demure, of lowly port,
> Or sprightly maiden of Love's court,
> In thy simplicity the sport
>   Of all temptations;
> A queen in crown of rubies dress'd;
> A starveling in a scanty vest;
> Are all, as seem to suit thee best,
>   Thy appellations.
>
> "A little cyclops, with one eye
> Staring to threaten and defy,—
> That thought comes next; and instantly
>   The freak is over;
> The shape will vanish, and, behold!
> A silver shield with boss of gold,
> That spreads itself, some fairy bold
>   In fight to cover.
>
> "I see thee glittering from afar,
> And there thou art a pretty star;
> Not quite so fair as many are
>   In heaven above thee!
> Yet like a star, with glittering crest,
> Self-pois'd in air, thou seem'st to rest:
> May peace come never to his breast
>   Who shall reprove thee!"

We may add that we know but of four references to the daisy in Shakspeare. In *Cymbeline*, act iv., scene 2:—

> "Let us
> Find out the prettiest daisied plot we can."

In *Love's Labour's Lost*, act v., scene 2 :—

> "Where daisies pied * and violets blue
> Do paint the meadows with delight."

Again, in *Hamlet*, act iv., scene 7 :—

> "There with fantastic garlands did she come
> Of crow-flowers, nettles, daisies, and long purples."

FIG. 33.

And, lastly, in *Hamlet*, act iv., scene 5 :—

> "There's a daisy; I would give you some violets, but they withered all when my father died."

In Milton there are but two allusions. In the Masque of *Comus* :—

> "By dimpled brook and fountain-brim,
> The wood-nymphs, deck'd with daisies trim,
> Their merry wakes and pastimes keep."

---

* "Pied," that is, vari-coloured, motley-coated.

And in *L'Allegro:* *—

> "Meadows trim with daisies pied."

The Tulip.

> "The pied windflowers and the tulip tall."
> —Shelley.

It is probable that, for the majority of floral amateurs, the name of the tulip is inseparable from a plant which, with the hyacinth and the lily, becomes, in the merry spring-time, the ornament of our gardens. Yet, towards the end of March, the observer will occasionally discover, in the woods and groves, the *wild tulip,*† the *Tulipa sylvestris* of Linnæus, which may, perhaps, be very properly taken for the type of a small tribe of the Liliaceæ. It is easily recognised by its flower, which resembles a large yellow campanula, slightly green on the exterior. Like all plants of the same family, it has but a single floral envelope or perianth, which may be either a corolla or a calyx as you will. The initiated protest and asseverate that it is a calyx; but the *profanum vulgus,* who compose the majority, will have it to be a corolla, on account of its colouring. To cut the knot, and please all parties, our beautiful floral envelope has been denominated a *petaloid perianth.*

The divisions of this perianth, six in number, may, in truth,

---

* See also Shelley's "Sensitive Plant," &c.
† This is common enough in Germany and France, especially in the vineyards, but very rare in England.

be considered as *petals;* they are detached down to the base, and full in proportion as the pistil is developed. The latter is composed of three stigmata, attached, without the intermediary of a stylus (sessile stigmata), to a free ovary (that is, an ovary not joined to the perianth), which, as it develops, forms a capsule with three angular projections marking so many lobes; each of these lobes includes a great number of compressed seeds. As in all the Liliaceæ, and in many other vegetable families, the stamens, six in number, are *hypogynous,*—that is to say, inserted at the base of the division of the perianth. The stem, nearly two feet in height, bears a single flower only: the leaves are lanceolate, like all of the family, and the root is formed of a bulb, with thin and brownish-coloured external *tunicæ.*

Is the wild tulip an original species, or only a degenerate variety of the cultivated tulip (*Tulipa Gesneriana*)? The question is one not very easily solved.

It is generally admitted that the cultivated tulip,—which everybody knows,—was introduced into Europe from the East, towards the middle of the sixteenth century. It is, at all events, certain that none of our older botanists speak of our wild tulip. Dodonnée himself refers to the Eastern tulip only, of which he was the first to give, in his "Historia Stirpium," a tolerable delineation.

A circumstance which would favour the belief that the tulip was imported from the East is the Oriental derivation of its name: *tulipa,* in Italian *tulipano,* comes to us, it is said, from the Turkish *tuliband,* or the Persian *dulbend,* whence is obtained, by corruption, *turban*, the characteristic head-gear of the Orientals. Thus, at bottom, *tulip* and *turban* are the same word, only altered in form.

Who does not know with what a glory of colours the skill of our horticulturists has succeeded in clothing the tulip?

Inasmuch as the cultivated species bears the distinctive addition of *Gesneriana*,—and of this species all existing tulips are but varieties,—we might reasonably suppose that Gesner, the celebrated Swiss naturalist (who died at Zurich, aged sixty-nine, in 1565), was the first to speak of it. But he makes no allusion to it in his "Historia Plantarum" (printed at Bâle in 1541); he only refers to it in his "Additions" to the works of Valerius Cordus, published in 1561.

We subjoin a literal translation of the words of Conrad Gesner:—

"In the year 1559, at the beginning of April, I saw at Augsburg, in the garden of F. H. Herwart, magistrate of that town, a plant whose seed had been brought from Constantinople, or, according to some, from Cappadocia. It was called *tulip*."

About the same epoch, this plant was cultivated at Vienna, in the gardens of some wealthy amateurs; whence several tulip-bulbs were afterwards sent into England.

This ornamental plant, whose splendour is of such brief duration, became, towards the middle of the sixteenth century, the object of a commercial speculation, which marks an epoch in horticultural annals. The towns of Amsterdam, Haarlem, Utrecht, Alkmar, Leyden, and Rotterdam, were the head-quarters of the new trade.

The years 1634 to 1637 marked its apogee, its culmination; it was the reign of the *tulipomania*,—a malady which, notwithstanding its severity, does not figure among our pathological nomenclatures. Bulbs of the variety called *Viceroy* were sold for 3000 florins (£235) each; and amateurs paid even as high as 5000 florins (£430) for the *Semper Augustus* variety! Those who had not the needful amount of ready money disposed of their goods, their cattle, and their furniture. And not only the horticulturists, but the seamen, and artisans, and servants, plunged headlong into this frantic gambling. Tulip bulbs were then as eagerly sought after as shares in the company of the Mississippi in the days of Law,—or in the South Sea Stocks, also set afloat by that ingenious financier.

But it was not so much a love of flowers as a lust of speculation which lay at the bottom of this famous mania. For example, a gentleman engaged a merchant to deliver, at the end of six months, a bulb worth 1000 florins. When the time came, the price of the bulb had either gone up or down, and the contractor paid only the difference; as for delivering the wares, neither party cared about it. It was, therefore, the exact equivalent of a speculation in the funds or in railway shares. The transactions took place on the public exchanges, as well as in coffee-houses, inns, and on the promenades. They originated a fertile crop of abuses, and to put an end to them the intervention of the Government was required.

However, we may cite several examples of distinguished men who have cherished a partiality for the tulip, in the better sense of the word. Among these was Justus Lipsius, the great philologist. In his garden he cultivated with his own hands, it is said, the rarest varieties, and his floricultural tastes were shared by two of his intimate friends, Dodonée (Diodati) and L'Écluse, the two most illustrious botanists of their time.

But all these details, however curious and interesting, do not teach us whether our wild tulip has sprung from the cultivated germs. As it is impossible to solve this problem experimentally, we are forced to be satisfied with a simple conjecture.

And, for our own part, we are strongly of opinion that the wild and cultivated tulips may, from their very origin, have co-existed independently of one another. And now to put forward a fact in support of this statement.

THE HELIOTROPE.

With the Heliotrope every lover of flowers is familiar; it is not less prized for its delicate fragrance than the tulip for its glowing colours. No doubt exists as to the country from which we have imported the cultivated heliotrope, nor as to the epoch when it was introduced: it came from Peru, whence the name given to it by Linnæus, *Heliotropium Peruvianium;* and was brought into Europe, in 1740, by Joseph de Jussieu.

Although not known in Europe above a hundred and thirty years, it is now an "old, familiar face" in every garden. Now, by the side of the cultivated species, a native of the New World, we can place a wild variety, indigenous to the Old World, common in our own country, and, indeed, in all the countries of temperate Europe; whence it has received the appellation of *Heliotropium Europæum*.

The European species, let me state, is in every respect similar to the Peruvian species, except that its flowers are inodorous and of a paler blue. Yet it was known before the discovery of America,—before the discovery of those regions from which we have obtained the cultivated heliotrope. Thus, the two varieties have existed contemporaneously, and have flourished independently of each other, from their very origin. Why should not such be the case with the wild and cultivated tulip?

The Anemones.

From our Spring posy the delicate Anemones must not be omitted. More than twenty species are cultivated in Great Britain, and I hardly know to which I would give the preference. They are called by that most unmeaning term, "florist's flowers," and from the attention bestowed upon them, the cultivated varieties have been greatly improved. But you and I, dear reader, will go forth into the "wild woods," and enjoy the rich gifts of nature untampered by horticultural science. It is towards the end of March that the wood anemone (*Anemone nemorosa*) begins to expand its graceful leaves and snow-white buds to the stray sunbeams that force their way through the embowering branches of stately elms and spreading beeches, and in April it has attained its full glory, contributing largely to the beauty and the show which then embellish the forest glade. Snow-white, and faint rose-red, and soft delicate lilac,—these are the prevailing hues of its tender petals.

It is said that the wood anemone never blossoms earlier than March 16, and never later than April 2. It opens out its loveliness to the sun about the same time as the swallow returns from the genial South to our land of pleasant verdure. Country children associate it with the appearance of the cuckoo, and call it the "cuckoo flower," but the "wandering voice" is later than the woodland blossom in its welcome to the spring.

Why is it called *Anemone?* Of course, the English name is derived from the Greek ἄνεμος, "wind;" but what connexion is there between the wind and the flower? Credulous old Pliny asserted that it never bloomed except when the wind blew. Some of our botanical writers explain that it shivers and bends before the winds of March and the breezes of April. Others remind us that though generally found in the shelter of the groves, it will thrive lustily in windy and exposed localities. But I suspect the true reason of the name is its peculiar sensitiveness to atmospheric changes. As a foreteller of the storm it is not less trustworthy than a barometer, never failing to fold up its exquisite petals when the winds are gathering over the distant hills.

Our plant is considered injurious food for cattle; and it was on account of its unwholesome properties, perhaps, that the Egyptians regarded it as an emblem of sickness; or the idea may have been suggested by its frail and feeble appearance.

The yellow wood anemone is a rare and beautiful variety, which I have sometimes met with among the chalky downs of Kent. Its botanical designation is *Anemone ranunculoides*.

A still richer species is the *Anemone pulsatilla*, or Pasque Flower Anemone; a silky downy plant, easily recognised by its blossom of glowing purple. The blue mountain anemone (*Anemone Apennina*) is only to be found, as its name indicates, on the bold rugged sides of lofty mountain-heights.

The Anemones belong to a very important order,—the *Ranunculaceæ*, or Crowfoot family,—which is divided into five sub-orders: 1. Clematidæ; 2. Anemoneæ; 3. Ranunculaceæ; 4. Helliboreæ; and 5. Actœæ, or Pœoniæ. [Linnæus distinguishes forty-one known genera, comprising a thousand species. There are nine British genera of Anemoneæ.

In Drayton's "Poly-Olbion" occurs a rich descriptive passage,—an exquisite "flower-piece,"—which, on account of its beauty, deserves to be better known, and more frequently quoted. The poet is enlarging upon the floral rites which were celebrated at the espousals of the rivers Thame and Isis, and sets before us a bright bevy of Nymphs and Naiads; engaged in twining "dainty chaplets" to deck the persons of the bride and bridegroom. The stalwart Thame, —so it seems to them,—should not be "dressed with flowers to gardens that belong," but with blossoms plucked from his own meads and pastures. As most of those selected are fit for a spring-time nosegay, we may well enrich our pages with quaint old Drayton's enumeration of them :—

> " The Primrose placing first, because that in the Spring
> It is the first appears ; then only flourishing ;
> The azured Harebell next with them they neatly mixt,
> T' allay whose luscious smell they Woodbine placed betwixt.
> Among those things of scent there prick they in the Lily,
> And near to that again her sister Daffodilly.
> To sort these flowers of show with others that were sweet,
> The Cowslip there they couch, and the Oxlip for her meet ;
> The Columbine amongst them they sparingly do set,
> The yellow King-cup, wrought in many a curious fret ;*
> And now and then among, of Eglantine a spray,
> By which again a course of Lady-mocks they lay ;
> The Crow-flower, and thereby the Clorra-flower they stick,
> The Daisy over all those sundry sweets so thick,
> As Nature doth herself to imitate her right ;
> Who seems in that her 'pearl' so greatly to delight,
> That every plain therewith she powdereth to behold.
> The crimson Darnel-flower, the Blue-bottle and gold,
> Which, though esteemed but weeds, yet, for their dainty hues,
> And for their scent, not ill, they for their purpose choose.
> Thus, having told you how the Bridegroom Thames was drest,
> I'll show you how the Bride, fair Isis, they invest."

Here the poet resorts to the garden for his decorative wreath, but is careful, as we shall see, to eschew " florist's flowers," and to select only our dear old favourites :—

> " The red, the dainty white, the gaudy Damask Rose,
> The brave Carnation, then, of sweet and sovereign power
> (So of his colour called, although a July flower),

---

* So we say, " fretted roof."

> With the other of his kind, the speckled and the pale;
> Then the odoriferous Pink that sends forth such a gale
> Of sweetness, yet in scents as various as in sorts;
> The purple Violet then the Pansy there supports;
> The Marigold above t' adorn the arched bar;
> The double Daisy, Thrift, the Button-Bachelor;
> Sweet William, Sops in Wine, the Campion, and to these
> Some Lavender they put, and Rosemary, and Bays;
> Sweet Marjoram with her like, sweet Basil rare for smell,
> With many a flower whose name were now too long to tell."

If our space permitted, we should like to gossip awhile about each of the flowers commemorated by our old poet, for to each attaches some legend, or romantic tradition, some rural observance, or sweet poetical association. But we must continue our researches, and they bring us now to the *Arum*.

The Arum.

To the French the Arum is commonly known as the Calf's foot (*Pied de veau*). It is a common enough plant, growing on the borders of the wood, and delighting especially in the shade of the hazel trees, but it bears not the slightest resemblance to the hoof of any quadruped whatsoever, unless, indeed, to a very fervid imagination there should be visible a shadowy similitude in its leaf.

And it is, in truth, asserted—but, not having the eye of faith, the editor cannot see any ground for the assertion—that its sagittate or arrow-headed leaves, marked by a strongly-defined mid-rib, bear a certain likeness to the "under bi-ungulated face" of the foot of a young ruminant. Appearing in the early days of spring, they contrast agreeably, by their shining verdure, with the colour of the dead leaves heaped up at the base of the hedgerows. Simultaneously with its leaves comes forth a curious organ,—rare in vegetables of temperate regions, common in the tropical palms, and characteristic of the family of the Aroidaceæ, to which our Arum belongs. This organ, rolled up in a coil or spiral, is named the *spathe*. It protects the flowers in their young state, and, as they are developed, gradually falls off. Its colour is a greenish yellow; at the summit it is sometimes streaked with purplish veins, and at the base it swells out in a globose fashion.

A small thermometer, introduced into the interior of the rolled-up spathe, indicates a rise of temperature equal to one or two degrees above that of the external atmosphere. Whence comes this difference? Because in the spathe is frequently found imprisoned another organ, the seat of the mystery of reproduction. This organ is a fleshy axis, on which are arranged the flowers in two distinct rings; the upper is occupied by the stamens, reduced to simple anthers (sessile stamens); observe the filamentous appendages—they are abortive ovaries. These same appendages also surmount the lower ring, where several rows are set of sessile ovaries; each ovary composed of a single lobe, containing a very small number of ovules, the majority of which miscarry as the ovaries become metamorphosed into bright red berries: these are the fruits which appear in autumn; they form a spike or ear of coral, each containing, ordinarily, a single seed. The flowers, as a consequence of this separation of the two sexes, are *monœcious;* the succulent axis which bears them is called a spadix.* On tearing open the spathe, our glance first rests upon the apex of the spadix, which has a club-like form, and is of a beautiful violet-red colour. The two rings of sexual organs have much less attraction for the profane; the lower ring, loaded with female flowers, is more prominent than the upper ring, which bears the male flowers.

The root of our Arum also deserves a particular examination. It is a white tubercular stock or stem, containing a quantity of fecula, mixed, as in the West Indian manioc, with an acrid poisonous principle which produces a burning painful heat in the throat. This injurious principle is destroyed by exposure to the fire, and by repeatedly boiling the plant in water. After being thus heated, there remains only the fecula, in the form of a white powder, which, in times of scarcity, supplies a very nutritious food. "I made use of it," says Bosc, "during the storms of the Revolution, when I had taken refuge in the solitudes of the forest of Montmorency. This plant is so abundant in this forest, and in many other localities, that, at the epoch I speak of, it would have ensured the subsistence of several thousands of men, if they had known its alimentary properties. I was seriously counting on the resources which it would place at my disposal, when the death of Robespierre relieved me from my difficulties." †

* When the spike bears numerous flowers, surrounded by a spathe, or sheathing bract, it is called a spadix.

† Bosc, a distinguished naturalist, died in 1828, aged 69. In the ministry of Roland, he accepted the delicate post of administrator of prisons; was proscribed after the terrible events of May 31, 1793; and lay concealed, along with Laréveillière-Lépaux, for several weeks in the forest of Montmorency.

Our *arum*, which we have taken as a type of the family of the Aroidaceæ, is called *maculatum*, or "spotted," in allusion to the white and violet spots with which its leaves are besprinkled.

Another, and not less interesting species, is the *Arum arisarum*. (See Fig. 34, *a*.) It loves to display its exquisite leafage on the rocks bordering the "sea-marge," and is found in profusion along almost the entire littoral of the Mediterranean. It is a precocious flower — making its appearance about the end of December, and flourishing until the beginning of Spring. The spathe, which in the *Arum maculatum* has all the aspect of an etiolated leaf, assumes, in the *Arum arisarum*, the tints of a corolla,—is of a beautiful warm red violet, streaked with white. The fleshy axis, which ought rather to be called gynandrous (both male and female) than a spadix, is of a red colour; naked in its upper portion, which terminates with a kind of apple. It would remind a drummer-boy of the formidable staff carried by his drum-major (see Fig. 34, *b*.); the stamens, reduced to the condition of bilobed anthers, are mounted around the central part; and the ovaries, less numerous than the stamens, occupy the base of the axis. Each monocular ovary is crowned by a sessile stigma, and each lobe contains a great

Fig. 34.—The *Arum arisarum*.

number of erect ovules. In the *Arum maculatum*, the number of ovules does not exceed six. Some botanists have laid hold of this characteristic as an excuse for withdrawing the Mediterranean species from the *arums*, and creating a new genus, *arisarum*. The variety we have just described is, in that case, denominated the *Arisarum vulgare*.

The ancients have mentioned numerous species of the *arum*. But it is a very difficult task to bring their nomenclature into any kind of agreement with the species described by modern botanists. However, we may, I think, regard the *arisarum* of Pliny and Dioscorides as positively identical with our *Arum arisarum*. But we are unable to admit that the *aron*, the *hepha*, the *dracunculus*, the *dracontium*, can be, as commentators represent, one and the same plant; still less can we admit that this plant is our *Arum maculatum*, which is very much rarer in the south than in the north and centre of Europe. In the solution of such problems as these, geographical botany is an element which must not be neglected. Unfortunately it has never been taken into account by the commentators on the great classical authorities.

Let me advance a simple proposition. Since the potatoe has become diseased, and the species tends to degenerate, may we not find a substitute for it,—at least, a partial one,— among our Aroids, and, notably, in the *Arum maculatum* ?

The Ranunculaceæ.

Let us return for a while to the order of Ranunculaceæ, of

which the Anemones have already furnished us with a specimen. Several very poisonous plants are members of this order; and, in truth, very few can be pronounced wholly innocent. I do not think there is much harm in the *Lesser Celandine*, however—the glossy, starry flower, yellow as a buttercup, with heart-shaped leaves, which Wordsworth has celebrated:—

> "Ere a leaf is on the bush,
> In the time before the thrush
> Has a thought about its nest,
> Thou wilt come with half a call,
> Spreading out thy glossy breast,
> Like a careless prodigal;
> Telling tales about the sun,
> When we've little warmth or none."

There cannot be much harm in it, for in the north of Europe the peasantry boil its leaves, and eat them as greens. It thrives in all parts of England, in green woods and meadows, and on wild furzy wastes and open commons; under leafy hedges, and even in the gay pastures, among the primroses and hepaticas. A number of small, grain-like tubers lie around it, close to the surface of the earth; whence it was a common saying in "the days of old" that this plant showered down wheat in its vicinity.

To the same order belongs the *Buttercup* (*Ranunculus bulbosus*), whose bulbous root procured for it from our forefathers the name of "St Anthony's turnip."

If the good saint ever partook of buttercup-corms, we do

not envy him his sensations; when boiled, they disorder the stomach, and if eaten raw, act as an emetic.

It was formerly thought, says a pleasing writer, that crowfoot (the buttercup is a species of crowfoot), mingled with the pasture, improved its nature, and that the butter yielded by cows which fed upon this mixture was of a superior quality. *Nous avons changé tout cela;* we are wiser now; and have discovered that cows carefully avoid eating buttercups, and that several kinds of crowfoot are even poisonous to cattle. On some pasture-lands, in those countries where the produce of the dairy receives particular attention, women and children are employed to destroy the crowfoot, which they do either by pulling up the root, or by plucking off the flower, and preventing it from dispersing its seed. The root of the buttercup is of a highly stimulating property if taken in an uncooked state, and its juice will occasion sneezing; but boiling deprives this, as well as many other vegetable productions, of its injurious properties. A similar effect is produced by drying it in the sun; wherefore the hay crop is not at all deteriorated by its acrid nature.

A very beautiful ornament of still pools and gently-flowing streams is the *Water-ranunculus* (*Ranunculus aqua atilis*), whose leaves vary according to the depth, or calmness, or swiftness of their watery habitat, and are thus adapted to permit the passage of water without suffering any injury from its force. The leaves on the surface have a round lobed shape; those immersed hang down in thin small fibres, which offer but little resistance to the current.

The RANUNCULACEÆ also include the *Black Hellebore*, or Christmas rose (*Helleborus niger*), one of our most splendid winter-garden decorations, whose juice the ancients considered a wonderful remedy for mental disorders. In whiteness it rivals the snow, which often accumulates around it, and the snow-drop, which is frequently bound up in the same wreath. It is called the *Black Hellebore*, to distinguish it from the two wild species which grow in our woods, its root being covered with a thick black skin.

The fragrant white *Clematis* must not be omitted; its starry drops are "things of beauty," which every true poetic eye will know how to appreciate. It is sometimes called "Traveller's Joy," and sometimes "Virgin's Bower;" either name is richly suggestive of pleasant fancies. Do you remember the beautiful picture in Keats's "Endymion," of the shady sacred retreat where Adonis lay and slumbered? The clematis was one of the precious flowers that adorned it:—

> "Above his head
> Four lily stalks did their white honours wed,
> To make a coronal, and round him grew
> All tendrils green, of every bloom and hue,
> Together intertwined and trammelled fresh;
> The vine of glossy sprout,—the ivy mesh,
> Shading its Ethiop berries,—and woodbine,
> .Of velvet leaves and bugle blooms divine;—
> Convolvulus in streakèd vases blush,
> The creeper mellowing for an autumn flush,—
> And *Virgin's Bower* trailing airily,
> With others of the sisterhood."

Finally, our order comprises the *Hepatica*, with its blue or pink blossoms and three-lobed leaves, which, from their fancied resemblance to the form of the liver, procured the plant its English name of liverwort; the *Flos Adonis*, or pheasant's eye,—the *goutte-de-sang* of the French,—so called because the ancients fabled that it sprang from the blood of Adonis, when wounded by the bear; the marsh marigold; the gay and vivacious larkspur; the deadly wolfsbane, or aconite, which secretes so potent a poison; and the aromatic love-in-a-mist, or French flower.*

B.—Animals.

Under the soft moss, under the stones, in all localities where mouldiness is easily developed, under the closed doors of cellars, you must certainly have more than once observed a tiny creature of the form of a horse-bean, of a gray leaden colour, and supplied with a considerable number of feet. This last characteristic will induce you immediately to abandon the idea that you have before you an insect.

Catch hold of it, and count its feet.

Well said; but it runs much more quickly than I would have suspected from its previously dilatory movements.

Because it knows that danger threatens it, instinct impels it to escape at its utmost speed. Do not be afraid to handle it; the poor creature can do you no harm.

* The seeds of the latter are used in the East, where they are more pungent than in our cold climates, instead of pepper. "They are thought to be the cummin alluded to in Scripture, where our Saviour reproved the Pharisees for their singularity in minor things, and their neglect of important duties."

Unable to escape, it counterfeits death, and remains perfectly immovable.

Now examine it. It has fourteen feet, symmetrically arranged in couples; their size perceptibly increases from the first to the last. When the animal is at rest, they are coiled inside, so as to form an angle whose opening faces the medial line. But here is something much more curious; its body, which does not possess the vestige of a wing, is also without those segments which would divide it visibly, as in the case of insects, into head, thorax, and abdomen; but is composed of rings, hard and scaly, like those of a shrimp.

Can it be a crustacean?

Yes; the animal you hold in your hand, and which everybody knew by the name of wood-louse long before our naturalists knew how to classify it, belongs to the great animal division of the *Articulata*, which, instead of having their skeleton *inside* the body, like the Vertebrates, have it externally. The Crustacea form a class of this division, to which also belong the Insects, the Arachnida, and the Myriapoda.

Let us continue to anatomise our crustacean.

In front of its first ring (a transversal segment) you see a little black head, with two lateral bead-like eyes, and a couple of antennæ. The latter are each composed of three joints, which are extremely mobile; their base is covered by the edges of the sloping head. The most conspicuous rings of the body are seven in number; their lateral borders are pointed in front, and rounded behind. But, if you look closely, you will see some other rings, a little less projecting than the former; they circumscribe the abdominal region, the belly, properly so called, in which the intestines are lodged. These rings, or abdominal segments, are six in number; but they have not all the same form. The one which occupies the tail is triangular, pointed, and surrounded by four (caudal) appendages. The three next segments, counting from the front to the rear, are prolonged laterally in a very marked manner; the two anterior, on the contrary, have no such distinction. As for the caudal appendages, the two outer ones are very strong, conical, and composed of two articulations, while the inner, situated above the former, are frail, cylindrical, and terminated by a tuft of hairs, whence issues a viscous liquid. (See Fig. 35, *a*.)

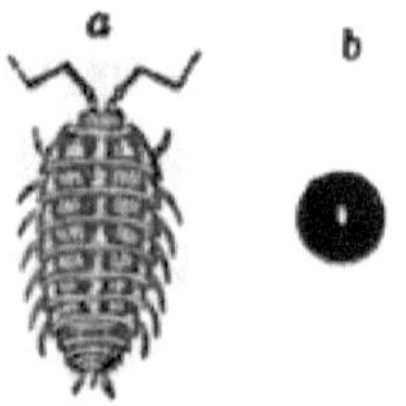

FIG. 35.
The Wood-Louse.

An enumeration of these characteristics is tedious, but necessary for the determination of the genus and the species. They belong to the *Oniscus asellus*, or common wood-louse. But why, you ask, why such a strange conjunction of names, —one Greek, ὀνίσκος, the other Latin, *asellus*? Both carry the same meaning: why not, then, have called our tiny crustacean an ass-ass (if such a compound be possible)? Why, neither close at hand nor at a distance, has it the slightest resemblance to an ass; and to say that we have only borrowed these names from the ancients, is neither an explanation nor a justification.

But we have not yet done with the wood-lice. Are these interesting little creatures (they *are* interesting, are they not?)

oviparous or viviparous? I defy you to show me anywhere a single wood-louse's egg. Have the patience to observe our crustaceans more nearly. Among the crowd, you will remark some—they are the females—with a kind of membraneous pouch underneath the body, stretching from the head to the fifth pair of legs. The pellicle which forms this pouch is so thin, and so transparent, that you can distinguish the eggs within it. These eggs, instead of being expelled for incubation, remain in the mother's pouch until they are hatched. At that felicitous moment, the membranous bag splits cross-wise, longitudinally and transversally, to permit the emergence of the young wood-lice. The latter are extremely small, and in form resemble nothing in the world so much as a little white line (Fig. 35, *b*). They differ from their parents only in having one pair of feet, and one ring less than they have. They undergo no metamorphoses. After their birth, the little ones, which have proportionally very large antennæ, do not immediately separate from their mother. By a wonderful act of forethought on the part of Nature,* they keep themselves concealed in the middle of the respiratory *laminæ*, which garnish the under part of the tail.

The specific characters of the *Oniscus asellus* are tolerably well defined. By its rings of dark gray, a little lighter at the edges, which form for it an articulated, glossy carapace, marked with white spots, longitudinally arranged; by the uniform pale gray colour of its belly and its legs, covered with scattered

---

* That is, the Creative Power which, in common parlance, we choose to call Nature.

hairs; and, particularly, by its habits, our wood-louse, which the Germans call cellar-louse (*Kellerlaus*), is distinguished from its kindred species, of which naturalists have made distinct genera. Thus, the *asellus* found generally under stones, which counterfeits death by rolling itself up in a ball like a hedgehog, and will rather suffer itself to be crushed than unfold, is the *Oniscus armadillo*, which some naturalists transform into the *Armadillo vulgaris*. (See Fig. 36.) This species prefers the solitude of the field to inhabited places. Its body is considerably expanded, and its rings do not terminate in a point on their lateral and posterior edges.

Another species, equally common underneath stones, has its head and tail covered over with granulations; its antennæ are composed of seven joints, of which the fourth and fifth are perceptibly situated lengthwise. This is the *Oniscus granulatus* of some entomologists; others have designated it the *Porcellio scaber*. Why not simplify the study of species?

Fig. 36. *Oniscus armadillo.*

The wood-lice seem to live upon decomposed vegetable matter. But in default of other food, they devour their own kind; in this respect resembling beings who are supposed to rank much higher in the animal hierarchy.

In the pharmacopœia of the ancients our wood-lice found a place. Reduced into powder, and mixed with various substances, they were prescribed as diuretic and aperient; but they were long ago abandoned in medicine.

THE DRAGON-FLIES (*Libellulæ*).

In walking along the banks of a river, you must frequently have seen hovering around you a cloud of insects, whom you would readily take to be butterflies, were not you arrested in your conjecture by the largeness of their head, the length of their body, the form of their vivid, diaphanous, gauze-like wings, and, generally,—which will most astonish you,—by their carnivorous instincts. You have about you and before, then, not butterflies, but Dragon-flies—the *Libellulæ* of naturalists. They are the *demoiselles*, or "ladies," of the French; so called, perhaps, in allusion to their airy and graceful flight.

Among these Libellulæ, one is called *Eleanora*. If she does not shine so brightly as the others—if her colours are less brilliant—she has, at least, the advantage of being so common that you can easily obtain a specimen.

But, first, let us pause to think of the strange dissimilarity in the names bestowed on the *Libellulæ* by the English and French respectively. They are the *Dragon-flies* of the former,—fierce, rapacious, formidable; the *Ladies* of the latter,—elegant, light, and radiant. Here we have a glimpse of national character. With the Frenchman, "appearance" counts for so much; with the Englishman, everything depends upon the "reality." Yet our English poets can appreciate their gay exterior. Moore speaks of them as—

> "Those bright things which have their dwelling
> Where the little streams are welling!"

Poor Clare, the Northamptonshire poet, correctly studied—

> "The great dragon-fly with gauzy wings,
> In gilded coat of purple, green, and brown,
> That on broad leaves of hazel basking clings."

And Mary Howitt has seen them—

> "Here and there they dart,
> And flush like gleams of green and azure light."

Beautiful as they are, they must be ranked among Nature's fiercest and most insatiable destroyers. They are the terror of the insect world. On this point we shall hereafter enlarge, but before I forget it, I would fain relate an anecdote, in illustration of their voracity, which I have read somewhere or other. A naturalist recounts with what interest he has often watched the proceedings of the dragon-fly. He has seen it, in a locality where white butterflies were numerous, dart down as a hawk upon a quarry, seize with its legs a firm hold of a butterfly, and carry it to a branch of an adjoining tree. In a moment one of the white wings would drop from the boughs, and then another would come wavering downwards, and so on, until all four had fallen; and the dragon-fly, after a short pause, would again dart forth in pursuit of a fresh victim. He never launched himself on his prey when on a perfect horizontal line with it; but took care to be either somewhat higher, or somewhat lower, so that he could seize it with his feet.

FIG. 37.—The Common Dragon-fly.

But now let us consider by what characteristics the reader is to recognise his Libellulæ.

The eyes are very large, the size of the insect considered; of a brown colour, and nearly joined together at the top of the head; in front of this point of junction a tiny vesicle is visible, carrying three *ocelli* (*i.e.*, little, simple, glossy eyes), two on each side, and the third on the anterior margin: the thorax is large, hairy, and composed of two yellow plates; the abdomen laterally depressed, in such a manner as to give great prominence to the medial line. This last and well-marked feature it is which has procured for the Eleanora the scientific name of *Libellula depressa*,—a name proposed by Linnæus, and unanimously adopted; such unanimity being a rare occurrence among naturalists, though, to parody a phrase of Sheridan's, when naturalists *do* agree, it is something wonderful!

In the male Libellulæ, the upper surface of the abdomen is bluish in hue, and covered as it were with an ashy dust, while in the female it is olive; in both sexes the first and last abdominal segment are of a deeper shade than the other segments. The feet are black, and bristle all over with stiff hairs; the thighs are of a brownish red. The two pairs of wings, each strongly reticulated, present, towards the extremity of the upper border, a black rectangular spot; their base, moreover, is edged with brown spots; the spots of the lower pair are triangular and larger than those of the upper pair, which are nearly linear. They form, to a certain extent, the reservoir of the liquid which nourishes and maintains the circulation of the network of the wings.

The Libellula, which Geoffroy calls (in his "Histoire des Insectes") the *Philinta*, is simply the male of the species we have just been describing. You may see them

> "Where water-lilies mount their snowy buds."

pursuing one another with an abrupt, jerking—I had almost said *staccato*—flight.

A species less common than the preceding, but closely resembling it, is the *Françoise* of Geoffroy, the *Libellula quadrimaculata* of Linnæus. It owes its descriptive or specific designation to the colours which diversify its wings. On the outer edge of each, two brown marginal spots are conspicuous: the first at the place where in the Eleanora (and other species) the black spot is found, and the second nearly in the centre of the external border, which, at this point, is considerably compressed; moreover, the lower wings are marked, beneath their yellow base, with a kind of triangular spot of blackish brown, finely reticulated with yellow. Externally, there is no difference of appearance between male and female, except that the abdomen of the latter is somewhat the larger.

Now we come to the *Sylvia*, the *Libellula cancellata* of Linnæus. Its eyes and thorax wear a greenish hue; the diaphanous wings are spotted with brown near the outer margin; the abdomen is of a bluish-gray; the extremity of the sixth segment, and the following segments, are wholly black.

The *Julia*, or *Libellula grandis* (Linn.), has been separated from the Libellulæ by the great entomologists, such as Fabricius and Latreille, and included in the genus *Aeshna* (Phœbus, what a name!), preserving the descriptive adjective *grandis*. What are the reasons put forward to justify this separation? Principally, the form of the abdomen, and the position of those little smooth and simple eyes, like tiny pearls or *pearlets*, which we call *ocelli*. In the Libellulæ, properly so called, the ocelli, three in number, are situated on either side and on the exterior margin of a kind of semi-triangular vesicle, and the abdomen is slightly depressed, and not unlike a club; while in the *Aeshnæ* the ocelli are placed on a simple keel-shaped transversal embossment, and the abdomen is narrow, elongated, and almost cylindrical.

The *Aeshna grandis* (or the "Julia") is one of the largest of the British species. Its head is large, and its eyes are of a brown colour shading into blue; the yellow thorax has two bright yellow bands or stripes obliquely painted on each side. The abdomen is of a reddish or even rusty brown; generally spotted with white and yellow at the top and bottom of each wing. This species haunts the vicinity of streams and "silent pools."

We come, in due order, to the *Aeshna forcipata*, or "Carolina," with its dirty yellow head, and its greenish-yellow thorax, the latter marked on each side with three oblique lines of black: the abdomen is black, and composed of segments laterally spotted with yellow.

The *Libellula*, which the illustrious Geoffroy designates "Louisa," is, according to modern entomologists, neither a *Libellula* nor a *Aeshna*, but an species of *Agarion*,—the *Agarion virgo* (*Libellula virgo*, Linn.) The Agarions are distinguishable from the Libellula and the Aeshna by their remarkably thin, filiform, and exceedingly elongated abdomen, and by the three ocelli arranged triangle-wise on the top of the head.

The "Louisa" (*Agarion virgo*) is by no means uncommon in England, and may be found along the upper course of the Thames, the Avon, and other rivers. It is easily recognised by its frail slender body, shining with metallic blue reflexes. There are numerous varieties, distinguished by their varieties of "light and shade." The spotless green-winged species is the "Ulrica" of Geoffroy. The two sexes are not alike. In the centre of their delicately-reticulated wings the males have a large bluish-brown spot, which is wanting in the females.

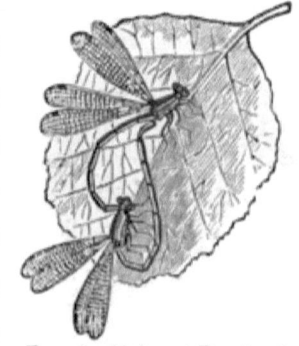

FIG. 38.—Male and Female of the *Agarion virgo*.

The genera *Libellula*, *Aeshna*, and *Agarion* compose the small family of the *Libellulites*—a family plainly and conspicuously characterised by the size of their head, and by the two pairs of diaphanous wings of almost equal dimensions (the posterior pair is a *little* shorter than the anterior), which, while the animal is at rest, are kept horizontally extended.

FIG. 39.—" A sunny pool, half-fringed with trees."

These are tne characteristics which present themselves to the observant eye at the first glance. Do you doubt me? Betake yourself, when

> " The bird is building in the tree,
> And the flower has opened to the bee,"—

betake yourself, I say, to any sunny pool, half-fringed with trees, or pleasant river-margin; go, armed with net and microscope, and, having secured a specimen of these terrors of the insect world, where

"The strong on weak, cunning on simple, prey,"

devote yourself to its patient examination.

It is only such an examination that can reveal to us some equally important, but less obvious features of the insect. But to catch a Dragon-fly is not always an easy task; the Libellulæ are very timorous, or else they are suspicious of the prowling naturalist who seeks what he can entrap. Their flight is livelier and swifter than even that of the butterfly: when disturbed in their repose, they fly away abruptly, their wings rustling or crackling like a sheet of parchment; if obstinately pursued, they grow irritated, and in their quick jerking movements exhibit all the rage of the Carnivora.

But can our Libellulæ be carnivorous? Most undoubtedly. To convince yourself of it, you have but to glance at their mouth, which is wholly unlike a butterfly's. What an arsenal of weapons adapted for seizing and crushing a victim! How strong are those saw-toothed scaly mandibles! How strong their auxiliaries, the jaws, which terminate in that dentated spring projection, furnished internally with ciliæ! Surely, such instruments testify to their ferocious instincts, and should induce our French neighbours to deny them the graceful name of "les demoiselles." What a libel on tender woman,—on man's "ministering angel!" Do but observe them. They do not rest upon the blossom to extract its nectared sweets;

in truth, they could not do so, for they are not furnished with a proboscis. Warlike as the Amazons,—the only portion of the female sex they can justly be said to resemble,—they hover in the air to pounce, like vultures, upon whatever insects may come within their reach; they quickly transfix, and as quickly devour them. If they love to fly about the pools, the marshes, and the streams, it is because they are sure of prey in these localities. And, in fact, they there encounter and devour an innumerable quantity of flies, moths, gnats, and the like.

But there is another reason why the Dragon-flies, obeying the secret impulse of nature, resort to the haunts we have been describing. They were their cradles or nurseries, and they become, in due time, the scenes of their espousals.

Before speaking of the singular metamorphosis they afterwards accomplish, we must touch lightly on the subject of the mode of reproduction of our brilliant *demoiselles*.

It is laid down as a law that, in the insect world, the males are invariably smaller or weaker than the females. Yet this law does not hold good with reference to the Libellulæ, whose males are, on the contrary, larger and stronger than their females. Man may lay down laws, and extort obedience to them, within his own domain; but nature laughs at human rules, and gives up her secrets only to the free thought, unshackled by the fetters of authority.

But why is the male Dragon-fly stronger than the female?

Because the former must make the first advances, and carry off his aërial companion to celebrate their bridal. For this purpose, he holds her tightly by the neck, and continues to fly in this way for some few minutes. At length, he perches himself on the branch of a willow, or the leaf of an aquatic plant, along with his companion.

The eggs laid by the female are oblong in shape, and sometimes united together in clusters: the female deposits them in the water, or on some water-floated leaf plant, shortly after their fecundation.

METAMORPHOSIS OF THE DRAGON-FLY.

" To-day I saw the Dragon-fly
 Come from the wells where he did lie ;
 An inner impulse rent the veil
 By his old husk : from head to tail
 Came out clear plates of sapphire mail.
 He dried his wings : like gauze they grew,
 Through croft and pasture wet with dew,
 A living flash of light he flew."
   —TENNYSON.

If in your walks abroad you should meet with a pool of turbid water, do not object to linger for a moment on its brink. Nature, as a reward for your trouble, will gratify you with an extraordinary surprise. Stir up the mud with a stick, or, better still, with your hand; and then, as the children say, you shall see—what you shall see!

Oh, what an ugly creature is struggling in this handful of slime! It looks like a large spider!

Nay, examine it more attentively; it can do you no harm.

Well, it is a singular animal; greenish in colour, bristling with hairs, and covered with mud. It is not a spider, for it has only six legs and these are exactly like the legs of insects. I can distinctly make out three joints (articulations) to each tarsus, which terminates in a simple hook. The belly is formed of regular segments; is rounded above, and flat underneath. What do I see? On the top, and nearly in the middle, each ring is armed with a spine, so that the row of spiny projections remind one of the back of a crocodile. Pray tell me, what is this curious creature?

Before I tell you, I would have you more thoroughly acquainted with it. Continue your examination.

The face has a strange expression; it looks—the insect, I mean—like a masked knight. Its mask, formed of a scaly substance and thoroughly compact, is composed of two pieces, which are separated from one another by a transversal suture: the upper, which is broader than it is high, we may call the vizor; it consists of two lobes, soldered together longitudinally; the lower, which is higher than it is broad, will be the chin-piece; it is triangular in shape, and its base rests against the vizor, while the summit is jointed or articulated with a support, which acts as a hinge when the mask is raised or lowered.

These movements, let me tell you, are voluntary; the animal raises its mask to arrest on their way the Infusoria and other animalcules on which it feeds. For crushing them, it is provided with strong mandibles, which you can detect by lowering the mask with a pin, or the point of a penknife. The eyes, which resemble little mammillary protuberances, are situated above and outside of the vizor. On the animal's back, where the belly joins the thorax, do you observe those four little sheath-like or scabbard-like tongues? The extremity of the body, the tail, is marked by three conspicuous triangular points, lying close to the opening, through which the water enters and issues, as if it were alternately sucked in and poured out by a piston. This, you must understand, is the respiratory apparatus. (See Fig. 40.)

FIG. 40.—Larva of the *Libellula depressa*.   FIG. 41.—Larva of the *Agarion virgo*.

Well, then, what is the name of this most singular creature, this masked knight?

It is the larva of the *Libellula depressa;* the dismal envelope whence will issue the gaudy Dragon-fly. Listen to a graphic description of the mask you are looking at :—

"Conceive your under-lip to be horny instead of fleshy, and to be elongated perpendicularly downwards, so as to wrap over your chin and extend to its bottom; that this elongation is there expanded into a triangular convex plate attached to it by a joint, so as to bend upwards again and fold over the face as high as the nose, concealing not only the chin and the first-mentioned elongation, but the mouth and part of the cheeks; conceive, moreover, that to the end of this last-mentioned plate are fixed two other convex ones, so broad as to cover the nose and temples; that these can open at pleasure transversely, like a pair of jaws, so as to expose the nose and mouth, and that their inner edges, where they meet, are cut into numerous short teeth, or spines, or armed with one or more sharp claws;—you will then have as accurate an idea as any powers of description can give you of the strange conformation of the under-lip of the larva of those insects, which conceals the mouth and face precisely as I have supposed a similar construction of *your* lip would do yours. When at rest, this mask applies closely to and covers the face; when they would make use of it, they unfold it like an arm, catch the prey at which they aim by means of the mandibuliform plates, and then partly refold it, so as to hold the prey to the mouth in the most convenient position for the operation of the two pairs of jaws with which they are provided."

And so the creature I have been examining is only a larva! How strange to compare it,—thick, ugly, unwieldy,—with the insect that issues from it, so aërial, so graceful, so light, so beautiful! The more I think of the contrast, the more it interests me.

This larva, however, has all the characteristics of a perfect insect, and I will wager that more than one observer has described it as such, and classified it among aquatic insects. Yet it is but a larva! And each species has its own special larva (see Figs. 40 and 41).

This seems to me as difficult to believe as if you told me that John, James, Peter,—in fact, all men,—were only temporary bodies from which more perfect beings would one day emerge. For the medium in which these masked creatures live,—the troubled and muddy medium for which they are adapted by their organisation, and in which they seem destined to live out their life,—differs wholly and absolutely from the aërial medium wherein, you tell me, they will one day rejoice.

Do you see, on the brink of the pond, those dried-up bodies? They are those of your aquatic insects, the mortal remains of Libellulites; they have rent open their vest, as you would do a garment which had become too narrow; the rent is very conspicuous along the back; and through the fissure issue the winged and aërial resuscitated Dragon-flies, summoned to live in a world differing so widely from their former one! Under the tongues were folded up the wings, and the mask that puzzled you so greatly is split open on the level of the sutures, so as to represent a T, wanting only a handle to reproduce one of those hieroglyphics ($\overset{\circ}{\text{T}}$) which are found so frequently on the Egyptian monuments, and which, according to some Egyptologists, signify *eternal life*.

And here, my friend and companion, we may take leave of

" The great Dragon-fly with gauzy wings."

BOOK III.

SUMMER.

Blessed, thrice blessed is the man with whom
The generous prodigality of nature,
The balm, the bliss, the beauty, and the bloom,
The bounteous providence in every feature,
Recall the good Creator to His creature,
Making all earth a fane, all heaven its dome! . . . .
   The sod's a cushion for *his* pious want,
And consecrated by the heaven within it,
   The sky-blue pool, a font;
Each cloud-capped mountain is a holy altar;
   An organ breathes in every grove,
   And the full heart's a psalter,
Rich in deep hymns of gratitude and love.
               —Thomas Hood.

The poetry of Earth is never dead. . . . .
A thing of beauty is a joy for ever.
               —Keats.

CHAPTER I.

WHAT MAY BE SEEN IN THE HEAVENS.

"And God said, Let there be light; and there was light."
—Genesis.

" A sound of song
Beneath the vault of Heaven is blown."
—Goethe.

THE unequal duration of Day and Night, the succession and regular return of the seasons, all the phenomena observable upon the earth, are but the effects of a cause which we must seek in the heavens. It is impossible to explain *them* unless we contemplate *it* on high, relegating our planet into the great chorus of the worlds, where it holds but a modest rank. Only, to perform this miracle, we must for a moment repress in ourselves the senses which deceive us by their exaggeration or appearances, and give free course to enlightened thought. It is by this means alone that we can succeed in fully demonstrating the close and perpetual relationship which exists between our planet and the other spheres composing what we call the Solar System.

Permit us here a parenthesis, or, shall we say, a digression?
The whole secret of science, the whole secret of human knowledge—in truth, the whole future of humanity—lies in these two words—*Enlightened Thought.*

And here, gentle reader, I solicit your assistance in endeavouring to elucidate a question which has a for a long time puzzled me. Why do we apply the word "light" to that which sets in motion the eye of the body, and to that which induces the operation of the eye of the mind,—to the singularly mysterious physical agent without whose intervention all the external world would be to man a dream or a void, as well as to the still more mysterious moral agent which illuminates the world within?

I anticipate your answer. "The light," you say, "which opens up to me the material world is a reality; the other is only an image."

And it is true that this is the solution which first presents itself to the mind. But the longer you reflect upon it, the more you will be inclined to agree with me that it is unsatisfactory. What, then, means this agreement among all peoples, past and present, to designate by the same words—as

Light	To illuminate
Obscurity	To obscure
Shadow	To overshadow
Morning	To darken

Day	To clear away
Twilight	To blind
Gloom	To cloud

—those actions, processes, or operations which take place or are produced in the outer world, physical or material, and in the inner world, intellectual or moral.

Were your pretended "image" the result of purely personal impressions,—all persons are not equally apt in apprehending those fictitious relations which are the food of poetry,—were it, in a word, no more than an individual conception, and, therefore, eminently variable, I should not hesitate to accept your opinion: we should simply be discussing what are called, in scholastic language, the individual sense (*sens propre*) and the imaginative sense (*sens figuré*). But there exists upon this point an unanimous and universal agreement: all languages, living or dead, attest it; it embraces the aggregate of the members of the human family.

This is the *first* point which induces me to doubt the legitimacy of the proposed explanation. But I am confirmed in my hesitation by numerous other facts.

Let us emerge from the domain of rhetoric to enter that of experimental physiology and psychology. Too powerful a light blinds the organs of seeing. No mortal eye, unless with the aid of artificial appliances, can gaze upon the sun, the great source of light. If the vision is to act clearly, we must proceed step by step. If from a very light apartment we pass rapidly, and without transition, into a darkened room, or *vice versa*, we feel temporarily blinded: the eye demands a short time to recover, as it were, from its astonishment. If we fix our gaze for one or two minutes on a star of the first magnitude, as, for example, Sirius, and afterwards turn away abruptly, the eye will for a while remain insensible to stars of a less intense splendour.

These are incontestable physiological facts, which anybody may easily verify for himself.

Well, in the psychological order facts exist which are in all respects analogous. Take one of those truths which the human thought, labouring generation after generation, has taken centuries to discover or elucidate; place it suddenly before an unprepared mind; however luminous may be this truth, it will be simply shadow and darkness to the mind we speak of; it will not comprehend an iota of it. Why? For the very obvious reason that it lies outside its sphere of ideas, in every respect comparable to the sphere of vision, beyond which and within which there is no more room for the sensorial impressions. There, as here, we must proceed gradually, and arrange the transitions so as to produce the desired results.

It would be easy for me to develop this parallel by other and still more remarkable facts; but what I have just said will suffice to show that the line of demarcation which philosophers have, upon principle, desired to trace between the physical and moral order, has turned the mind aside from many fertile fields of research and speculation. Let us cite an example. The sun is the visible centre of light, heat, organic life; in fine, of all the movements of our material world. Yet it is but a relative focus, since the sun, with its planetary train, revolves, probably, in company with other suns or systems, around a centre as yet unknown; and as there is no reason why we should pause in this cycloidal progression, this second centre or focus of systems may revolve around a third, the third around a fourth, and so on. Thus we shall have an indeterminate series of relative centres; for the term does not exist of which we can say,—there is the beginning, or here is the end of the series.

We do not meet with the absolute in the material, any more than in the intellectual world. Truth, by its power of attraction, sets in motion all the wheels of our understanding; we seek after it eagerly, in the doubt which torments us, in the obscurity which surrounds us; we all feel the need of being enlightened by it, and warmed, and revivified; we all are in need of belief, and, at the same time, of possessing—let there be no illusion in this respect—a certainty or demonstration of what we believe.

But the truth which we think our own does not leave the mind at rest; a slight effort suffices, in fact, to teach us that the truth we accept depends upon another and remoter truth, and that the world of thought is thus carried onward in an interminable series of relative truths; unless we find it more convenient to pause here at a primary cause, as elsewhere at a primordial centre, which we may ever identify with the primary cause. But is this truth, supposed to be final, capable of satisfying equally every mind? With this question, dear reader, I close my parenthesis.

To understand clearly the variable duration of the day at all points of the terrestrial surface, let us so place ourselves as to see our planet distinctly in front. I need hardly say that this invitation is not addressed to the visual organs of the body, but to the "mind's eye, Horatio." With the former, fixed as we are on the surface of the globe, we can only see the shadows of objects standing out in relief upon it, and still, with the sun at our back, can perceive but a comparatively insignificant portion. With the second, on the contrary, we may detach ourselves from the earth, may mount afar into space, may plant ourselves on some aërial height, whence we may embrace, at a single glance, all the illuminated hemisphere of our globe: we shall have but a single shadow on the prospect,—that which our brightened planet casts behind it, and which, in four-and-twenty hours, traverses the entire surface of the globe, producing light wherever it passes.

Having completed our preparations, let us note what we perceive with our eye thus disjoined from our body,—an eye equally well adapted to fix every moving object, and to see at any distance bodies extremely small, or bodies extremely large. What a marvellous eye! And it is for us to develop its power, and freely to increase its range. What an arduous task!

The luminous envelope, the photosphere of the sun, simultaneously darts its rays in all directions, with an intensity which diminishes with the square of the distance.

Let us follow those which direct their course towards our planet.

So long as the rays do not come into collision, or strike against resistant matter, they neither warm nor enlighten. Confirming the Newtonian hypothesis of emission, they travel straight as arrows through the icy space, the shadowy ocean in which the waves of the terrestrial atmosphere are lost. But the moment they encounter a material obstacle, the rays partly penetrate it, and are partly thrown back, in such a manner as to form a series of undulations like those which the falling of a stone into a pond produces. As they recede from these repeated shocks, the rays come under that theory of undulation which Huygens promoted.*

Let us trace the rays of light back to their origin. These encounter first the globes nearest to the sun: the almost imperceptible asteroids, which, on account of the solar splendour, can scarcely be detected, and which, as yet, have received no baptismal names. Yonder beams illuminate other revolving globes,—Mercury, Venus, the Earth. If we here arrest our gaze, we are influenced by a vague and mournful recollection; the Earth was our place of sojourn at an epoch when thought, formerly shackled, had become free. It is thus that a seed, re

---

* No one, I think, has ever before attempted to reconcile, in this way, the two principal theories which have been put forth on the propagation of light.

moved from the stem which nourished it, wanders afar to diffuse and perfect its species. Our planet, of an ochreous yellow, relieved with green and white, possesses no special privileges; it shares in the ponderated movements of the spheres; it is neither the largest, nor the smallest, nor the heaviest, nor the lightest, nor the nearest to, nor the farthest from, the sun. And shall Earth alone, of all the planets, nourish that kind of "thinking seed" which we name the human race? It seems improbable.

But let us return to the great orb of light. It illuminates exactly one half of our earth; the other half lies in shade. And as the earth rotates upon its own axis, every point of its surface is necessarily exposed to the action of the solar rays. This action varies in duration and intensity.

All this I was taught, when I was still at school on our revolving planet. I remember, too, as a lesson learned by heart, that, under the Equator, or, more exactly, in 0° 0' latitude, as well as at the Poles, or under 90° latitude; the duration of the day is equal to that of the night, with this difference, that, while under the Equator, a day of twelve hours alternates invariably with a night of the same duration, at the Poles a day of six months succeeds continually to a night of six months. I also recollect that, at a given moment, namely, at the spring and autumn equinoxes, the duration of the day, over all the terrestrial surface, is equal to that of the night, just as under the Equator; and that, after these two epochs, under the intermediary latitudes, between the Equator and the Poles, the length of the days and their corresponding nights varies according to the seasons; that, in our northern hemisphere, after the spring equinox, the days increase while the nights diminish, to such an extent that, at the summer solstice,—21-22 June,—they attain their *maximum* of length (and the nights their *minimum*), the reverse taking place during the period that elapses between the autumn equinox and the winter solstice.

I remember that I learned these data when I was living rooted to or anchored upon the earth; but the explanation which my masters gave me was not so clear as I could have wished. Their considerations on the declensions of the sun, on the obliquity of the ecliptic, on the necessity of exactly reducing the earth to a simple point in relation to the distance of the stars, to the end that the phenomena occurring (*rapportés*) on parallel planes would be nearly identical with those observed from the centre of the globe, or from a point situated on its surface; all these fine things, which demanded a certain faculty of geometrical intuition, left a curious vagueness of idea upon my mind. I accepted them, under the influence of authority, as beyond discussion, but I was by no means satisfied whenever I wished to ascertain their foundation.

But now—in the regions of space—everything grows simple before the mind, in which, apparently, all my power of thought is concentrated.

Behold the illuminated atmosphere; it reminds me vividly

of the disc of the full moon. Ah, what exquisite iris colours! They mark the meeting-points of the bright with the obscured hemisphere: a circular line carried through all these points would exactly separate day and night. On the one side, motion and life and glow; on the other, silence and shade and calm. This line *moves*, carrying with it in its movement day and night, the illuminated and the darkened hemisphere; it moves from east to west, so that the bright hemisphere and the shadowy one, whose union forms what may be called the *photo-adumbrated* sphere, revolves, in four-and-twenty hours, round an axis which coincides at this moment—the 21st day of March, according to the "terrestrial worms,"—with the axis of the earth rotating on itself in an inverse direction, that is to say, from west to east. Let us note this coincidence: it is remarkable; inasmuch as, at the equinoxes, the terrestrial equator divides exactly the illuminated and the obscured hemisphere into two equal parts, one of which is situated to the north, the other to the south, and their line of separation coincides with a meridian circle.

But I see another, and much slower movement, very clearly defined. The axis of rotation of the *photo-adumbrated sphere* does not remain parallel with the terrestrial axis of rotation; it retires from it little by little, so as to form with it an angle which attains its maximum at the summer solstice (21–22 June); afterwards, returning upon itself, it coincides anew with the terrestrial axis of rotation (the autumn equinox), to make an angle in the contrary direction, whose maximum, of the same value as the former, corresponds to the winter solstice (21-22 December). This movement, which is annual, complicates itself with the diurnal. It is rendered visible by the displacement of the polar shroud of snow: the portion which, at the moment of the spring equinox, belonged, in the northern hemisphere, to the obscured hemisphere, moves onward, as a result of the inclination of its axis, to become an integral part of the illuminated hemisphere; while, at the same time, the portion which, in the southern hemisphere, belonged to the illuminated hemisphere, moves onward to become an integral part of the darkened hemisphere. Owing to this displacement, the sun shows itself for six consecutive months above the horizon, for the North Pole; at the spring equinox it begins to rise, at the summer solstice it attains its maximum elevation, and from the summer solstice it begins to decline. Thus, then, we have in reality *a day six months long*, of which the morning and the evening are the two equinoxes, its noon the summer solstice.

In the same periods an exactly opposite order of things prevails in the southern hemisphere.

Everything, even to the minutest detail, is in this way very clearly explained. Two facts—like the touches of a painter's brush—suffice to impress the whole upon the mind, namely:—

That, first, one half of the terrestrial surface is constantly illuminated by the sun, while the other half remains in darkness;

That, secondly, the rotation of the earth upon its axis produces day and night, by carrying from east to west the illuminated hemisphere, always diametrically opposite to the darkened hemisphere.

This being thoroughly understood, let us place ourselves in the equatorial plane, so as to embrace at a single glance a quarter of the illuminated and a quarter of the darkened hemisphere. If the equator of the photo-adumbrated sphere perpetually coincided with the terrestrial equator,—if, in other words, the earth, in revolving round the sun, invariably occupied the plane of the Equator, which, when prolonged, would pass through the centre of the illuminating orb,—the diurnal rotation would not cease to divide equally the light and the darkness over the earth's surface, as is shown in Fig. 42 *a*,

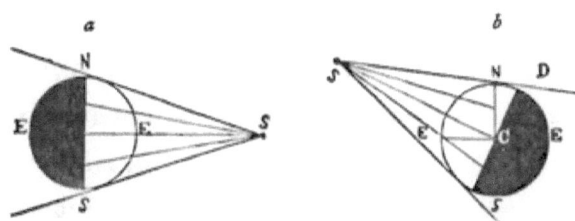

FIG. 42.—S, the Sun; EE, the Equator; N, North Pole; S, South Pole.

where S indicates the sun, EE the terrestrial equator, and N S the extremities (or North and South Poles) of the axis which divides the globe into an illuminated and a darkened half. This phenomenon of coincidence exists; but only for a very brief period, and is only repeated twice a year,—that is, at the equinoxes. At all other times, the equator of the photo-adumbrated sphere, in whose plane the orb of light is situated, passes sometimes above and sometimes below the terrestrial equator.

But this alternate movement of northern and southern declination has its limits; it stops at 23° 27′ 30″ on either side of the Equator; this, too, is the maximum of the distance which the imaginary axis of the photo-adumbrated sphere retires from the axis of terrestrial rotation; it is at the same time the value of the obliquity of the ecliptic, or of the inclined plane which the earth traverses in its annual movement of revolution.

Fig. 42 *b*, which represents the photo-adumbrated sphere at the summer solstice, will enable the reader to comprehend with the utmost facility the six months' day of the North Pole, coinciding with the six months' night of the South Pole; for the triangle N C D indicates the amount by which the illuminated moiety increases in the northern hemisphere between the spring equinox and the summer solstice, the amount being equal to that by which the adumbrated moiety overspreads in a contrary direction the South Pole in the southern hemisphere. The superfluous quantity of the photo-adumbrated sphere is *nil* at the apex of the two opposite triangles, or in the equinoctial region. And then, in effect, both day and night are always twelve hours long. Starting from the equinoctial line, we see how easy it is to calculate for each locality the variable dimensions of the arc which the sun, in its apparent course, traces above the horizon.

To see very distinctly the portion of the illuminated hemisphere, which, passing beyond the North Pole, forms a luminous course on the darkened moiety of our globe, I have but to place myself at midnight, on the 21st of June, in the prolongation of the terrestrial equator; in the same way, to see the corresponding portion of the darkened hemisphere, which advances beyond the South Pole to invade the illuminated, I have but to occupy at noon, on the 21st of June, a point of the same equatorial prolongation. Six months later, the same spectacle will be presented, inversely, on the 21st of December, in the southern hemisphere. See Fig. 43 *a*, where the pole N indicates the six months' day of the northern hemisphere (from the spring to the autumn equinox); while in Fig. 43 *b*, the pole S indicates the six months' night of the northern hemisphere (during the same period).

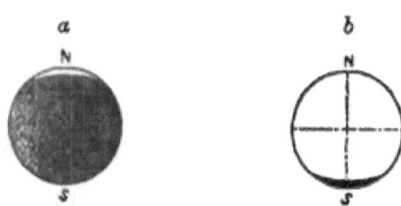

FIG. 43.—The Six Months' Day and Night.

Now, no effort of the imagination is required to understand why the inhabitants of the northern hemisphere enjoy summer while those of the southern are enduring winter; why it is "blossoming spring" to the former when it is "purple autumn" to the latter, and *vice versa*. Equally easy is it to comprehend why, after the equinox, *day*, or the duration of the sun above the horizon, gradually diminishes in one hemisphere and increases in the other; why, in summer, and in both hemispheres, the longest days alternate with the shortest nights, and in winter, the longest nights with the shortest days. It will not be more difficult to explain the cause of the prevailing cold in the polar zones, despite the prolonged sojourn of the sun above the horizon for a great part of the year. Observe how obliquely the solar rays are directed towards yonder shrouds of ice and snow:

how can they warm them? They nearly all vanish into space. (See Fig. 42.)

Finally, there is not a phenomenon, even to that of dawn and twilight, which cannot, on these principles, be very fully and clearly explained. I have indicated, in a preceding paragraph, the rainbow-glories of colour noticeable on the line of demarcation between the illuminated and the darkened hemispheres. They are wanting where the rays of light strike vertically or nearly vertically. It is this circumstance which explains why, in the intertropical regions, the crepuscular phenomena are nearly null; why the sun, so to speak, sets and rises abruptly, like a taper which we extinguish or re-kindle. These iris-gleams increase, on the other hand, in intensity, in proportion as we recede from tropical regions: the red touches the horizon, while the violet blends with the azure of the sky; between these two extremes, which are always very clearly marked, are arranged in less perceptible fashion, and in the order of refraction, the other colours of the rainbow.

What time and labour does it not require for the mind to disengage, to free itself from the fetters and incumbrances of sensorial appearances, the illusions of the senses, and to rise sufficiently high to seize at a glance all the dynamics of the world!

It is this faculty, however, which distinguishes the intellect from the imagination.

That he may abandon himself to the enjoyment of those pleasures which, like Dead Sea apples, crumble to ashes on his lips, the fool puts aside all mental toil, and disregards the shortness of his time,—ignores the brief period allowed for the development of the understanding. But, at least, let Imagination abstain from substituting its idle dreams for the assured results which can be only the reward of reason, conscientiousness, and labour! Unfortunately, here as elsewhere, it is *vox clamantis*, and we preach in the desert!

CHAPTER II.

WHAT MAY BE SEEN ON THE EARTH.

> "Now the shining meads
> Do boast the pansy, lily, and the rose,
> And every flower doth laugh as zephyr blows."
> —Ben Jonson.

THE Flower seems to have been created expressly to say to men:—"Listen! Those things which most attract your glance are but subordinate, and the principal escape you."

That the warning is true, all history attests. It is only, so to speak, from yesterday that the discovery of the sex of plants is to be dated; the tiny organs occupying the centre of the flower having always appeared so insignificant that they had passed, for some thousands of years, completely unnoticed. The eye of the spectator was caught by the calyx and the corolla; these envelopes, though of secondary importance so far as the reproduction of the vegetable is concerned, seemed to eyes dazzled by their glowing colours the true flower,—in fact, the *entire* flower. Science, which is the slow elaboration of thought matured by the study of objects of no human origin, has completely swept aside this premature judgment.

The Perianth.

We have already, and more than once, employed the word *perianth** to designate the calyx or corolla, whether taken separately or together. In the former case, the *perianth* is *simple;* in the latter, it is *double*. A more appropriate word could not be made use of. It is derived from the Greek περί, around, and ἄνθος, flower; and literally signifies, "floral envelope." Simple or double, this envelope is the metamorphosis of several leaves, never of a solitary one, inserted upon planes so closely brought together that they seem confounded. Observe, in fact, how the leaves tend to efface their intervals on the blossom-bearing spray; they draw towards each other, they are apparently in eager haste to accomplish their destined transformation. What eloquence there is in this simple language of nature!

The Calyx.—The outermost whorl, or verticle, of the flower

---

* The term *perianth* is usually confined to the flowers of Endogens, whatever colours they present, whether green, as in asparagus, or coloured, as in tulip. Some use the term as a general one, and restrict the use of *perigone* to cases where a pistil is present, not applying it to unisexual flowers, in which stamens only are produced."—*Professor Balfour*, "*Manual of Botany*," p. 169.

is called the *calyx*. And why? Out of a notion altogether incorrect. It is true that this foliaceous envelope may often assume the shape of a cup (in Latin, *calix*), and hence that the name has about it a semi-poetical air. But this only occurs when the calyx is composed of a *single* leaf, which has procured for it the special designations of *monosepalous*, *gamopetalous*, and *monophyllous*,—three different words expressing one and the same thing! The violet and the primrose are examples of a monophyllous, monosepalous, or gamopetalous calyx.

I see, dear reader, that you are puzzled by the word *sepal*. Certainly you would look for it in vain in any classical dictionary; it is neither Greek nor Latin. It was only invented, scarce a century ago, by a Swiss botanist, whose works have chiefly remained in manuscript,—by Necker, brother of the celebrated minister of Louis XVI., and uncle of the illustrious Madame de Staël. Let me explain the circumstance which determined, I suspect, the choice of this fanciful word,—a word belonging to no language but that of modern botanists.

The botanists of antiquity called the coloured leaflets of the corolla, *petals*. In this they were doubly right; for, first, they are, in reality, nothing but metamorphosed leaves; second, the word *petal* (in Greek, πέταλον) signified "a leaf" as early as the days of Homer, who, when speaking of the nightingale, says, like a keen observer of nature, that this bird, on the return of spring, sings—

> "Couched among the thick leaves of the grove."
> (Δενδρέων ἐν πετάλοισι καθεζομένη πυκινοῖσιν.) *

The word *petal* was preserved by Tournefort, handed down by Linnæus and the two De Jussieus, and afterwards adopted by all botanists with the signification given to it by the ancients. Now, as the calyx may also consist of leaflets, which are generally green, Necker conceived the idea of applying to them the same term, after substituting an *s* for the initial letter *p*. Thus was created the word *sepal*. The innovation, I must point out, was not unanimously adopted. Many botanists continued to use the words "calicinal leaflets," introduced by Linnæus; others, though they adopted the innovation, protested against it.

But leaving the *word*, let us return to the *thing*.

The calyx consists originally of several leaflets. Is the monophyllous or monosepalous calyx a transformation due to the junction of the primitive leaflets? Observation replies in the affirmative.

In the formation of junctures or adhesions nature proceeds from beneath to above. Our language proceeds inversely to nature: we speak of a lobed, dentated, or partite calyx, as if it were primarily monophyllous, and its more or less profound divisions (indicated by the words "lobed," "dentated," "partite") were but consecutive results, produced from above to below.

The truth is, that the calicinal divisions, which we call *lobes*, *lacinias*, and the like, are but the tops of leaflets united

* Homer, "Odyssey," Book xix., line 520.

at their base. The monophyllous calyx (formed of one piece) is, therefore, simply the result of a more or less complete union of the leaflets composing, properly speaking, the calicinal whorl. This whorl is originally *polyphyllous;* that is to say, formed of several distinct parts. If it were, in the first place, *monophyllous*, it would be impossible to understand how its divisions are made from top to bottom, since nature, in its developments, proceeds from bottom to top. In the final analysis, then, it is an error to consider the calyx as a cup, primarily formed of a single piece.

Grew, an English botanist of the eighteenth century, seems to have been the first who made use of the word *calyx*. " I call a calyx," he says,* "the external portion of the flower, which enfolds the others, whether it be all in one piece, as in the violets, or divided, as in the roses."

If we wish to conform to the truth, as brought before us by nature, we must revolutionize our terminology. Instead of speaking of bipartite, tripartite, quadripartite, or of bilobed and trilobed calices,—terms all signifying that the monophyllous calyx is cloven more or less deeply from top to bottom, we must say that the calyx in such and such a species has its leaflets united at the base, or one third, fourth, or half of its height; the *polyphyllous* or *polysepalous* calyx will be that whose leaflets remain detached, as was the case in the monophyllous or monosepalous calyx. This language, recom-

* Grew, " Anatomy of Plants," p. 147.

mended by the authority of Auguste Saint-Hilaire, would be more precise and exact: therefore, it will not be very quickly adopted. One would say that the human mind condemns itself to pass through the purgatory of what is false and complex, before resolving to adopt the simple and the true.

If we admit the theory according to which all the organs of the vegetable are the result of a metamorphosis of the leaf, we shall ask what place is to be given to the calyx in the series of these transformations?—Answer: The calyx is a foliaceous transformation, intermediary between the bracts and the corolla.

It is particularly in the study of the calyx that the attentive eye is struck by those proteiform movements in which nature makes sport of our absolute rules.

For example: in the *Berberis vulgaris*, the young calicinal leaflets have less resemblance to the bracts than to the petals of the corolla, and hence they have received the name of *petaloid sepals*. In other flowers this morphogenic wavering inclines towards the bracts rather than towards the corolla. We hesitate, therefore, whether we must give the name of *calyx* or *bracts* to the three under leaflets which are visible beneath the petaloid envelope of the *Anemone nemorosa*, or wood anemone. The calyx of the Celandine (*Ranunculus ficaria*), which also flowers in spring, is exactly like a whorl or involucre formed by the union of the bracts.

A similar embarrassment takes place when the calyx, in its metamorphosis, inclines too visibly in the direction of the corolla. Thus, in the *Polygala vulgaris*,—a little, vivacious, and very abundant plant,—the two inner leaflets of the calyx are not only larger than the three other outer leaves, but they are coloured like the petals, and become, towards the close of their flowering time, membranous, herbaceous, and marked with three strong veins: they resemble the wings of a butterfly, and have been called wings.

These peculiarities are useful in the distinction of certain species, which, at bottom, are simply varieties. Thus, in the *Polygala Austriaca* (*Polygala amara* of Keoch, *Polygala uliginosa* of Reichenbach),—a plant with small white or bluish leaves, which is sometimes met with on the borders of peaty swamps,—the central vein or ridge of the wings is simple, and never anastomoses with the lateral veins; while, in the *Polygala vulgaris*, as well as in the *Polygala depressa* and *Polygala amarella*, the vein is ramified, and anastomoses more or less widely with the laterals.

But since we are upon this subject, why should we not seize the opportunity of familiarising ourselves, under the form of a digression, with the little family of the Polygalaceæ? But no; we will adjourn the episode, since it would cut the thread of our discourse upon the calyx, our *calicology*.

Some calices there are, which, by their colouring, approximate so closely to the second floral envelope, that one is always tempted to call them corollas.

Such are :—The red calyx of the fuchsia ;

The yellow calyx of the furze (*Ulex Europæus*), and of a kind of hellebore (*Helleborus hyemalis*);

The rosy calyx of the Christmas rose (*Helleborus niger*);

The blue calyx of the larkspur (*Delphinium Ajacis*); and the Napel aconite (*Aconitum Napellus*).

In the *Aquilegia vulgaris,* and in the *Trollius Europæus,* the calyx, by the form and colouring of its leaflets, is confounded with the second whorl so completely, that Linnæus gave it the name of *corolla*.

Nevertheless, in the midst of these waverings, which lead us to mistake the calicinal leaflets sometimes for bracts, sometimes for petals, we recognise perfectly the foliaceous type. Independently of its colour, which is generally green, the calyx has the same organisation as the leaf; we find in it the same tracheæ and the same stomata, the same glands and the same hairs; the veins and ramifications are also the same; and, in more than one instance, the calicinal leaflet resumes the character of a veritable leaf. Look, for example, at the five leaflets, united so strongly at the base but so free at the top, arranged in the form of a quincunx, of the hundred-leaved rose. The two external, enlarged, and lanceolate pieces are garnished on the right and left, and often at the point, with tiny foliaceous appendages, which, in every respect, imitate the composite leaf that carries the slender stem. And if we move aside the external or bearded parts of the calyx, we see that the internal bear less and less resemblance to a leaf. Thus, the part which comes next is *semi-bearded*; that is to say, it is furnished with foliaceous appendages only on one side; and the two upper pieces are *beardless*, that is, reduced to the dilated central vein.

It was these metamorphic forms of the free portion of the calicinal foliola (united below) of the rose, which originated a well-known enigma, conveyed in the following Latin distich :—

" Quique sumus fratres, unus barbatus et alter,
Imberbes duo, sum semi-berbes duo, sum semi-berbes ego."

("We are brothers, both bearded, two beardless; I am two half-bearded, and I myself am half-bearded.")

They are specially noticeable in a variety of the Rose of Bengal, in which all the petals seem to be transformed into calicinal leaves. (Fig 44.)

The part of the calyx formed by the union of the sepals is called the *tube:* it is invariably the lower part. The upper portion, where the sepals are free, is the *limb.*

Throughout the vegetable kingdom you will not find a calyx in which the union of the sepals takes place at the top.

This time, at all events, we have found—what is exclusively rare in nature—a rule without an exception.

Generally, it is almost impossible to disunite, without rending, the foliola composing the tube of the calyx, their union is so complete. This circumstance prevented the first observers from accurately apprehending the composition and true development of the calyx. There are cases, nevertheless, in which Nature — the coquette! — suffers herself to be surprised, if her lover have patience. As an example we shall cite the monophyllous calyx of the Œnotheræ.

FIG. 44.—Rose of Bengal.

Let us take the species known as *Œnothera biennis*. It belongs, with the fuchsia, circæa, trapa, and others, to the Evening Primrose family, or Onograceæ.

Its pale yellow blossoms are unfolded during the hush of evening-time in almost every garden, shedding abroad on the breeze its delicate but delicious odour. Its petals open in a remarkable manner. The calyx, as we shall see, has small hooks attached to its upper extremity, by which it holds the flower together before expansion. The calicinal divisions gradually unfold at the lower part, and reveal the yellow flower, which remains awhile closed at the upper parts of the hooks. The flower then suddenly opens half-way, when it stops; afterwards completing its expansion gradually, and finally opening with a loud noise.

This curious plant is of American origin, and was unknown in our country until 1674, when it was introduced by some French floriculturists.

It opens generally at about six or seven o'clock in the evening.

And this statement induces me to digress. Where can I better introduce to the reader's notice a Floral Dial? It is not so complete as it might be made if I had space to enlarge upon the subject. My object, however, is simply to *suggest;* and this brief allusion to the hours at which flowers fold and unfold may induce the reader to study in more detail a very pleasant branch of botanical science. He will find full particulars in Mr Loudon's excellent " Encyclopædia of Gardening."

It is generally stated that the first Floral Dial, or clock, which showed the time by the opening or shutting up of blossoms throughout the day, was a fancy or invention of the great Swedish naturalist, Linnæus. But there is a distinct allusion to this poetical measurement of the "fleeting hours" in Marvell's poem on "The Garden:"—

> " How well the skilful gardener drew
> By flowers and herbs this dial new!
> Where, from above, the milder sun
> Does through a fragrant zodiac run;
> And, as it works, the industrious bee
> Computes its time as well as we.
> How could such sweet and wholesome hours
> Be reckoned but with herbs and flowers?"

Whether the idea first occurred to Englishman or Swede, poet or botanist, matters but little; it is a graceful, a suggestive, a beautiful idea, and might well be reproduced in some of our large public gardens.

"'Twas a lovely thought to mark the hours,
　As they floated in light away,
By the opening and the folding flowers,
　As they laugh to the summer's day.

"Thus had each moment its own rich hue,
　And its graceful cup and bell,
In whose coloured vase might sleep the dew,
　Like a pearl in an ocean shell."

FLORAL DIAL.

TIME AT WHICH THE FOLLOWING FLOWERS FOLD AND UNFOLD.

	Open. H. M. P.M.	Close H. M A.M
Goat's Beard (Lat. syn. *Tragopogon luteum*),	9.10	3.5
Late-flowering Dandelion (*Leontodon serotinum*),	12.1	4.0
Hawkweed (*Picris echioides*),	12.0	4.5
Alpine Hawk's-beard (*Crepis Alpina*),	12.0	4.5
Wild Succory (*Cichorium intybus*),	7.0	5.0
Naked-stalked Poppy (*Papaver nudicaule,*)	7.8	5.0
Copper-coloured Day-lily (*Hemerocallis fulva*),	11.12	5.0
Smooth Sow-thistle (*Sonchus lævis*),	12.0	5.0
Blue-flowered Sow-thistle (*Sonchus Alpinus*),	4.5	5.6
Field Bindweed (*Convolvulus arvensis*),	10.0	5.6
Common Nipplewort (*Lapsana communis*),	4.5	6.7
Spotted Cat's-ear (*Hypochæris maculata*),	5.0	7.0
White Water-lily (*Nymphæa alba*),	10.0	7.0
Garden Lettuce (*Lactuca sativa*),	3.4	7.0
African Marigold (*Tagetes erecta*),	2.0	8.0
Mouse-ear Hawkweed (*Hieracium piloscella*),	2.0	8.0
Proliferous Pink (*Dianthus proliferus*),	1.0	8.0
Field Marigold (*Calendula arvensis*),	3.0	9.0
Purple Sandwort (*Arenaria purpurea*),	2.3	9.10
Creeping Mallow (*Malva Caroliniana*),	12.1	9.10
Chickweed (*Stellaria media*),	9.10	9.10

After this long digression, we return to our Evening Primrose.

Its large yellow flowers are disposed in clusters at the

top of a stem often twenty inches long. The Pythagorean tetrad (*i.e.*, the number 4 and its double) predominates in all its organs: 4 stigmata crowning a filiform stylus; quadrangular capsule with 4 polyspermous lobes; opening at top by the separation of 4 valves; twice 4 stamens; 4 petals on a large, emarginated limb; 4 sepals. These are united at the base, but not so as to prevent the observer from distinguishing their number.

The general terms "regular" and "irregular," applied to the calyx, as to every other organ, require to be employed with considerable reserve. The delicate shades, which ought to separate regularity from irregularity, are often so inappreciable that it is almost impossible to say where one begins and the other ends. See, for example, the Labiatæ. In most genera and species of this family, the two lips, one of which consists of two and the other of three foliola, bring out very completely the inequality of the calyx. But there are also Labiatæ, the inequality of whose sepals completely effaces the character of the irregular bilabiated calyx.

In certain inflorescences, where the flowers comprising them are very close together, as, for example, in the capitules of the Synantheraceæ, the free upper portions of the calyx may take the most irregular forms; as, sometimes, a tuft, simple or feathery; sometimes, membranous or scarious spangles; and, sometimes, bristles of greater or less stiffness. What elements of the calyx do these transformations represent? The veins, and notably the midrib of the limb of the sepals, united underneath.

The free foliola of the *polyphyllous* calyx may vary in form, like the caulinary leaves whence they issue by way of metamorphosis; they may be oval, elliptical, linear, &c. Yet none have ever been observed of a heart-shape (*cordiform*).

Certain foliola of the polyphyllous calyx affect fantastic outlines. In the *Delphinium*, the upper sepal is prolonged into a spur. In the aconites it is hollow like a helmet. The spur of the calyx of the monk's-hood (*Capucine*) is the result of the united prolongation of these foliola. The buckler (*Scutellum*), from which is named the *Scutellaria*, a genus of Labiatæ, is a demi-orbicular boss formed below the inferior lip of the calyx.

FIG. 45.—The Henbane.

The union of the calicinal foliola sometimes forms a conical calyx, as in the *Silena conica*, and sometimes a cup-shaped calyx, as in the orange; sometimes, moreover, an urceolate calyx, as in the henbane (*Hyosciamus niger*). These forms may vary singularly. The calyx of the black alder (*Rhamnus frangula*) is shaped like a top; that of the haricot (*Phaseolus vulgaris*), like a bell; that of the tobacco-plant, or the *Mollucella spinosa*, is infundibuliform (or funnel-shaped).

The observer is sometimes embarrassed in deciding to which whorl he should refer the foliola he is examining. Thus, the tiny foliola which, in the strawberry and the potentilla, alternate with others and larger ones, are stipules rather than sepals. Ought the sepals of the calyx in the Malvaceæ to be assimilated in like manner to the stipules? It is difficult to reply to this question satisfactorily. Take, for example, the *Hibiscus Syriacus*, an ornamental shrub, better known by the name of the garden-hemp. The inner calyx, or calyx properly so called, of this Malvacea has five sepals, while the outer calyx, or calicule, has twelve. Now, a leaf cannot have more than two stipules, one on each side. For an outer calyx, then, the proper number of foliola is ten, not twelve. To look upon the second calyx as a "supernumerary development," would be to hazard a supposition contrary to the unity of plan of the floral organs.

The calyx, like the corolla, is not an absolutely indispensable organ. Sometimes, therefore, it is *caducous*—that is, falls off before the flower expands,—as in poppies; sometimes, *persistent*, or remains after flowering,—as in roses and the majority of plants. In some cases it is persistent only until after the act of fecundation, but this act accomplished, it falls with the corolla in most of the Cruciferæ and Ranunculaceæ. This is a *deciduous* calyx.

The "caducity" and "persistency" of floral envelopes furnish some valuable characteristics for the distinction of species. Thus, two closely-allied Cruciferæ, the *Alyssum calicinum*, so common in spring upon stony soils, and the *Alyssum montanum*, can only be distinguished from one another by the fact that the calyx of the former is persistent, of the latter caducous. It is true that the flowers of the *Alyssum calicinum* are of a yellow which easily passes into white, while those of the *Alyssum montanum* are of a beautiful permanent yellow. But this latter distinction is not so good as the former.

The persistent calyx sometimes assumes a considerable increase of very common appearance. For example, take the *Physalis alkekengi*, a member of the Nightshade family or Solanaceæ. The red bladder-like accrescence surrounding the scarlet fruit is the calyx, which, after flowering, has grown much larger than it was before. And these bright flowers which resemble large strawberries, and abound on the borders of meadow-paths, if you look at them closely, you find to be the accrescent calices of the *Trifolium fragiferum*. The very word *fragiferum* reminds us of the strawberry.

The calyx may change in consistency and texture in proportion as the ovary, to which it adheres, changes into fruit. The fleshy pulpy substance of the apple, and, in general, of the fruits of the Pomaceæ, is simply an excessive development of the calicinal tube united all around the ovary, and recognisable in the pips, imprisoned, towards the centre, in horny lobes. In other plants, as the flower develops into fruit, the calyx becomes woody: such is the case with the Water Chestnut (*Trappa natans*).

Finally, the calyx may even contribute to the dissemination

of the seed. We may cite, as an example, a Brazilian species of Urticaceæ, which Saint-Hilaire named the *Elasticaria*. The fleshy and cylindrical parts of the calyx are curved inwards, and thus defend, as one might do with one's bended fingers, the young fruit until it is completely developed; as soon as the fruit is ripe, they spring up erect, and launch it to a distance.

The Corolla.—If from the circumference we proceed to the centre of the flower, the calyx being the first, the *corolla* will be its second envelope.

If, on the contrary, we proceed from the centre to the circumference, the corolla will form the fourth whorl; the pistil (consisting of stigmata, stylus, and ovary) being the first; the nectariferous disc (often wanting) the second; and the stamens the third whorl. Remarkable for its varied tints, the corolla, to indifferent or ignorant eyes, seems the entire flower.

A black or blackish colour is exceedingly rare. Out of 300 vegetable species which compose the flora of Central Europe, there are not six with blackish or even grayish flowers. No hypothesis has yet been put forward to explain this markworthy rarity.

Species with a yellow corolla are the most numerous, forming more than a sixth of the European flora: then come, in their order of frequency, species with green, white, red, and blue flowers; the white increasing in number as we approach the Pole.

Dividing the flora into twenty parts, we may ascribe to each colour, and its various tints, the following proportion:—

Yellow,	.	6.0	Red,	.	3.50
Green,	.	4.50	Blue,	.	1.50
White,	.	4.0	Black,	.	0.50

The analogy between the parts or *petals* of the corolla and the leaf, is perhaps not quite so striking as between the leaf and the *sepals* of the calyx. The phrase "rose-leaves" is an expression consecrated by immemorial usage. Why not prefer the term "corollary leaves or leaflets (*foliola*)" to that of petals?

The corollary leaves, or petals, are organised like true leaves. They have the same system of venation; their *lamina* correspond to the "limb," or "blade;" and their *unguis*, or "claw," to the "petiole," or stalk. (See Fig. 46, *a*.) The upper margin of petals is frequently more obtuse than the overspreading margin of the blade of a leaf, which, in most cases, is pointed. Non-*unguiculate* petals represent the *sessile* leaves. (Fig. 46, *b*.) Their form is much more varied than that of the calicinal foliola, which are never unguiculate or petiolated.

FIG. 46, *a*.—Unguis of the Corolla.
FIG. 46, *b*.—A Sessile Petal.

The petal is defined as *regular* when its two halves, folded one upon another at the midrib, exactly cover each other; in the opposite case it is called *irregular*. In certain species, the petals are furnished with characteristic appendages. But observe, these appendages, which generally affect the form of a spur, have no character of generality. Thus, for example, in the violet, a single petal is prolonged into a spur below its point of attachment; in larkspur, and the other *Delphiniums*, there are two which terminate in the same manner; in the *Aquilegia vulgaris*, all the petals are *calcarate* (*calcar*, a spur).

According as the veins of the petal proceed in a straight or curved direction, its limb may be flat, or concave, or hollowed like a boat—*i.e., cymbiform* (*cymba*, a boat), or *naviculate* (*navis*, a ship), or like a spoon, *cochleariform* (*cochleare*, a spoon). When the spur is very short, as in Antirrhinum and Valerian, the corolla or petal is termed *gibbous* (*gibbus*, a swelling), or *saccate* at the base.

If a petal continue narrow, so as to seem formed by the prolongation of the claw, it is called *linear*; if the limb be prolonged below, so as to form two lobes, it is *cordate*, as in *Genista caudicans;* or if the lobes be acute, it may be *sagittate* or *hastate*.

The *number* of petals varies from two to twelve, and more. A corolla with a single petal, *unipetalous*, which we must not confound with the *monopetalous* corolla, is a monstrosity, created by defective development; the other petals or foliola are abortive. A corolla with two petals, or *dipetalous*, as in *Circæa Lutetiana*, is rare. A *tripetalous* corolla occurs only when the calyx has likewise three foliola. But in this instance opinions are divided: the majority will not admit more than a single floral envelope,—a perianth of six foliola, of which three, herbaceous and internal, alternate with three petaloid and external.

Many of our aquatic plants may be quoted as examples: such as the *Butomus umbellatus*, or flowering rush; the little frog-bit, or *Hydrocharis morsus ranæ*; and the water plantain, *Alisma plantago*.

The *tetrapetalous* (or four-petalled) corolla is usually arranged like a cross, and is much more frequent than the *dipetalous* or *tripetalous;* for examples, we need only refer to the large and important family of the Cruciferæ.

The number five (*pentapetalous*) is still more common; but we meet with it in other organs besides petals, and it seems particularly characteristic of the vegetable kingdom.

Thus, all the Umbelliferæ have five sepals, five petals, and five stamens; in all the Crassulaceæ, the number five applies, not only to the sepals and the petals, as well as to the stamens, but also to the carpels which compose the ovary.

Corollas with six, eight, nine, ten, or twelve petals are relatively rarer; and when the petals become so numerous that we cannot count them, we have to deal with transformations of stamens into petals, with those monstrosities of cultivation which we call double flowers, *flores pleni*, where all the male organs have disappeared,—flowers wholly unfit for fructification.

The petals of the corolla are not always free. Like those of the calyx, they may be attached to one another by their edges, but this union invariably takes place, as in the former, from bottom to top. Therefore, we never see any petals united at the top, and disengaged at the bottom (see Fig. 47, *n*.) But the reader must take careful note that this invariable characteristic is not peculiar only to the floral envelopes; it is not met with in the stamens, for these may be united, either by their anthers, as in the whole family of the Synantheræ, or by their filaments, as in the Leguminosæ. And what we have said of the stamens applies also to the parts constituting the pistil. This radical difference between the perianth and the true reproductive organs ought, from the beginning, to have fixed the attention of botanists on the centripetal and centrifugal metamorphosis of the leaf, which we have spoken of in "The Circle of the Year."

FIG. 47, *a.*—Natural junction of Petal.
FIG. 47, *b.*—Unnatural (and impossible) junction.

In many plants we are permitted to follow step by step, as it were, the union of the petals, and their definitive transformation into what is called the *monopetalous* corolla. The term *monopetaloid* ought then to be rejected, if we are to believe that the *monopetalous* corolla is the result of the metamorphosis of a single leaf. The word *gamopetalous*, or, rather, *gamophyllous*, is preferable. As we have remarked in reference to the calyx, we shall here repeat that the expressions *bilobed, trilobed, quadrilobed,* or *bipartite, tripartite, quadripartite* corollas, are radically vicious, because they are the consequence of a false point of view, according to which the *monopetalous* corolla will be simply a single metamorphosed leaf, susceptible of being more or less deeply divided from top to bottom.

The polypetalous corolla, as well as the gamophyllous corolla, may be regular or irregular, according as the foliola are equally or unequally united. But here again we must be careful not to lay down too absolute rules. Examine, for instance, the gamophyllous corolla of the gentians; their two lobes are very unequal, and yet the corolla is regular: five very large lobes alternate with five very small, in such wise that each of the latter is situated between two larger lobes. Each division of the corolla of the periwinkle is irregular, and yet the aggregate of their divisions is regular: these are all of the same form and the same size.

In the gamophyllous corolla we are able to discern, through the different forms it assumes, the form of the parts which compose the union. Thus, for example, the *tubulate* corolla supposes the pre-existence of unguiculate petals.

Inequality of union produces the *bilabiate* corolla, which is invariably tubular. This is the case in the natural family of the Labiatæ. The upper lip is composed of two petals, and the lower of three. The parts of the upper are sometimes so closely joined that they appear to be but one, as in the *Lamium;* and the lower often becomes quadrilobed by the division of the middle petal, as in the *Stæcha*.

In the labiated corolla, the mouth of the tube is open, while in the *personate* or *masked* (*persona*, a mask) it is closed by the pressure of the lower lip against the upper; as in Snapdragon and Frogsmouth, the projection of the lower lip being called the *palate*. By this feature the family of the Scrophulariaceæ is easily distinguished from that of the Labiatæ.

In some corollas, the two lips are hollowed out in a very singular fashion, as in the Calceolaria; assuming a "slipper-like appearance," similar to what takes place in the labellum of certain orchids,—to wit, the *Cypripedium*. These *calceolate* (*calceolus*, a slipper) corollas may be looked upon as consisting of two slipper-like lips.

The forms of the bilabiate, tubular, and ligulose florets, of the capitula of the Synantheræ, are likewise due to simple differences of union. The floscular capitulum comprises the tubular florets, and the semi-floscular capitulum, the ligulate florets; the radiate capitulum consists of florets ligulate or bilabiate at the circumference, and of tubular florets over the rest of the receptacle. By considering the capitula, as the vulgar do, to be flowers, Tournefort introduced considerable confusion into the nomenclature of the Synantheræ.

SUMMER FLOWERS.

"Dawn, gentle flower,
 From the morning earth!
 We will gaze and wonder
 At thy wondrous birth!

"Bloom, gentle flower,
 Lover of the light,
 Sought by wind and shower,
 Fondled by the night!" *

---

* Barry Cornwall.

THE PRUNELLA, OR SELF-HEAL.

In your summer-walks, dear reader,—summer-walks through green lanes rejoicing in the shade of over-arching elms, or

FIG. 48.—" Rejoicing in the shade of over-arching elms."

along woodland glades, carpeted with odorous turf,—you must frequently have met with an herbaceous plant, whose purple-blue flowers, arranged in regular succession, form the prettiest coloured cones imaginable at the extremity of the stem and branches. To this plant, the self-heal, we shall return immediately. Perhaps you have passed it by somewhat indifferently, for we pay little heed to common things, and on the threshold of woods, and in their winding avenues, the self-heal is very common. The celebrated German botanist, Bock (or Tragus) bestowed upon it, two centuries before the epoch of Linnæus, the name of *Prunella vulgaris.* The specific appellation, *vulgaris,* is here employed very appropriately, but we should commit a grave error if we supposed every species qualified as *vulgaris* to be "common." For example, the *Lysimachia vulgaris,* a species of Primulaceæ, is far from being found everywhere.

Pray, take the trouble to pick one lowly specimen; being specially careful to take up the whole plant, stem, root, and branch. Lying along the ground, it seems larger than it really is. Its root is a creeper; at the level of the insection of the leaves some small shoots project, the fibrous radicles which compel the lower part of the stem to crawl like the bugle, *Ajuga reptans.*

Does the prunella belong to the family of Labiatæ, like the bugle?

FIG. 49.—The Prunella.

See for yourself. The stem is quadrangular; the branches and leaves composed of two lips. The stamens are four in number, two of which are longer than the others; finally, by means of a lens, you can easily distinguish, at the very bottom of the calyx, four tiny seeds (a *tetrachænium*) grouped around the style. These features indicate that our plant belongs, in effect, like the bugle, to the Labiatæ family.

But mark the difference. In the bugle, as in all the species of the same genus (*Ajuga*), as well as in all the *Teucriums*—of which wild sage (*Teucrium scorodonia*) is the most widely-diffused type—in all the Labiatæ, the corolla is apparently *unilabiate*,—that is to say, the upper lip is so shortened that only the lower is prominently visible. This is not the case with our self-heal: it is distinctly bilabiate. The upper lip of the corolla here forms a positive *hood*, sufficiently ample to protect the didynamous stamens (two long and two short), as in Fig. 50, *a*; the lower lip is three-lobed, and the central lobe is largest of the three. By separating the two lips, you can see the two short stamens fixed to the base of the lower one, and the two long attached to the central part of the upper. (See Fig. 50, *b*.)

Fig. 50.—The Lips of the *Prunella vulgaris*.

Let us pursue the analysis of the flowery cone you hold in your hand.

The least practical eye is immediately struck by the arrangement of the parts and the variety of the colours. To recognise these things more thoroughly, please to cast your glances alternately from the top to the base, and from the base to the apex of its terminal flower. A little below the base you will see a pair of opposite, entire, and sinuous leaves, with shorter stalks than any of the others. The base is defined by two opposite, whitish, and nearly triangular leaves, with green points. The top of the floral spike is likewise marked by a couple of bracts; but these are much smaller, and red-coloured, like the two leaves of the calyx. The interval is occupied by bracts, which diminish in size from the base to the top of the spike; on a level with each pair six flowers are inserted, three for each bract.

The flowers, thus arranged by whorls, present some interesting peculiarities. The lower and upper show only their reddish calices; the middle, for the most part, display both a calyx and a corolla, varying from blue to pale-rose, which gives the plant a very peculiar appearance. In the under flowers, the corolla has already fallen; by separating the lips of the calyx, you may catch sight of the tetrachænium, that is, the four-seeded fruit, which is developed at the bottom of the tube. In the upper flowers, the corolla is not yet expanded. It resembles a small deep-coloured globe; you may say an eye, a bull's-eye, which, from the depths of the calyx, regards you with a piercing glance. Hence, perhaps, the French name for this plant, *prunelle*, an eye.

We often meet with a variety of self-heal with a white corolla, green calyx, and pinnatifid leaves, a variety of which some botanists have erroneously made a separate species, under the name of *Prunella alba*. It is equally wrong, in our opinion, to convert the large-flowered variety into a distinct species, by taking as its specific character the *lateral cleft of the upper lip of the caly overlapping the middle cleft;* for this same characteristic is found in many individuals of the common species. The *dentiform appendage* which the two longest stamens exhibit at the top of their filaments is also an uncertain feature; you must have recourse to your magnifying-glass to see if this appendage is *obtuse* and very short, as in the large-flowered prunella, or *sharp*, as in the common species. As for the size of the corolla, it is, in fact, very marked; but, as a characteristic, is wholly insufficient. The creation of the varieties *Pinnatifida, Laciniifolia,* and *Integrifolia* is no better justified. For it is no rare thing to see on the same stalk, at different heights, pinnatifid, whole, and laciniate leaves.

The prunella is remarkable for the long hairs which garnish the calyx, and, principally, the edges of the bracts. Examined in the microscope, they assume the form of tiny, pointed bamboos; the knots bulge out a little, and the intervals are punctuated.

The botanists of the sixteenth and seventeenth centuries are by no means sparing in their eulogiums on the marvellous virtues of our flower, which, by the way, Bock (Tragus) was the first to figure with tolerable accuracy.[*]

Tournefort thus dwells upon its medicinal properties :—

"It forms an ingredient in arquebusade water and vulnerary potions. It is ordered in possets and broths, in apozèmes for the spitting of blood, dysentery, hæmorrhages, and the like. It has also been used for ulcers in the mouth, and a remedy against headaches; after being mixed with rose-oil and vinegar, the temples were bathed with it."

[*] Tragus, "Historia Stirpium," p. 310 (ed. 1552).

In the Pharmacopœia of to-day, however, it finds no place.

The Scutellaria.

At the first glance, the Scutellaria has no resemblance to the prunella. Yet the classificators have united these plants in one small tribe, under the name of *Scutellarinaceæ*. These are the characters which they give to them: Lower or anterior stamens longer than the superior or posterior; calyx closed at maturity by the approximation of the two lips. The latter character is not nearly so marked in the prunella as in the scutellaria.

The two commonest species of scutellaria in England are the *Scutellaria galericulata* and *Scutellaria minor*. They do not inhabit the same localities. The former, which is at the same time the commonest, grows on the river-banks, and especially delights in the mould accumulated in the hollow trunks of old willows. It is easily known by its tender blue corolla, but especially by its calyx, which, after the fall of the corolla, develops itself in a singular manner. If you compress its sides, it will open so as to disclose, at the bottom of its

Fig. 51.—*Scutellaria galericulata.*

throat, its seeds, which are white, red, or brown, according to their degrees of maturity. (Fig. 52, *a*.) Now look at these two *jaws:* the upper resembles a small helmet (Lat. *galericula*), or, if you prefer it, a judge's cap. As for the lower, it has exactly the shape of a shield (Lat. *scutum*),—whence its name, *scutellaria*. (Fig. 52, *b*.) Thus, the emblems of military and judicial rank are found united in the calyx of our pretty labiate.

FIG. 52.—Calyx of the Scutellaria.

The second species (*Scutellaria minor*), rarer than the former, is met with on the banks of ponds and in damp woodland paths. It attracts your gaze by its tiny caps or helmets: the moment you see it, you exclaim, "That's a scutellaria!" More diminutive in all its parts than its congener, it is also distinguished by its whole leaves (they are crenelated or dentate in the *Scutellaria galericulata*), by its soft, rose-hued corolla, with brown lips coquettishly pointed with red, and by its hairy calyx.

The *Scutellaria Columnæ* * is very rare. It may be recognised by its erect stems and flowers of a bright violet hue, arranged in terminal spikes and garnished with bracts; while the flowers are axillary, and form no spike, in the two species above described.

The *Scutellariæ* were first described with accuracy and classified by Linnæus, who included them among his *Didynamia*, a class of vegetables distinguished by the unequal length of the stamens.

THE FORGET-ME-NOT.

"Ye field flowers! the gardens eclipse you, 'tis true;
Yet, wildlings of nature, I dote upon you,
For ye waft me to summers of old,

---

* Fabius Columna (Fabio Colonna), an Italian man of science, who died in 1650, at the age of eighty-three.

When the earth beamed around me with fairy delight,
And daisies and buttercups gladdened my sight,
   Like treasures of silver and gold.

FIG. 53.—"A loved little island, far seen in the lake."

"Even now, what affections the violet awakes !
   What loved little islands, far seen in the lakes,
      Can the wild water-lily restore !
   What landscapes I read in the primrose's looks !
   What pictures of pebbles and minnowy brooks
      In the vetches that tangle the shore !" *

* Thomas Campbell.

"How beautiful," says Miss Pratt, in one of her agreeable little books,*—" how beautiful are the little islands of the stream, edged with the tall white meadow sweet, which sends its perfume far up over the green lands that lie around, and contrasts with the deep blue colour of the purple loose-strife! The willow herb, or codlins-and-cream, as the children call it, grows in perfection there; and there, too, bloom the little yellow water-flag, and the vetches, and the rich water-lily, which, seated on its round leaf, seems to swim over the crystal stream. The water-plantain, with its numerous small pink blossoms, grows in thick clusters quite down in the water, mingling with the white flowers and large spear-shaped leaves of the arrow-head, or half shading the large cup of the yellow water-lily. Then, too, the blue-eyed forget-me-not covers the little isles in such abundance that many of them well deserve the name of azure islands. The water-rat hides among the flowers, nibbling with much glee at the arrow-head, or rushing out from under its broad green leaves; and the water-fowl, followed by her young, sails across the stream in all the stateliness of matron dignity; and the little meek-eyed daisy grows beside the yellow velvet flower of the silver-weed, or the blue blossoms and succulent leaves of the brook-lime."

The *true* Forget-me-not, *Myosotis palustris*, is invariably found in marshy localities or on the banks of streams; but the Meadow Scorpion-grass, *Myosotis arvensis*, is frequently mistaken for it. The "genuine article" has a bright blue blossom, much smaller than, but in shape something resem-

* Anne Pratt, "Flowers and their Associations" (ed. 1846).

bling, the primrose; in its bract it has a drop of gold, and on each segment of the coloured portion of the flower is a small streak or fleck of white.

Both the true forget-me-not and the false belong to the Borage family, or Boraginaceæ, which includes sixty-seven known genera, and nearly nine hundred species.

It is said that after the battle of Waterloo, a remarkable number of forget-me-nots sprung up all over the fatal field. The circumstance might well be made the theme of a poet's lay, were it not for a suspicion that the little blue flowers belonged to the *Myosotis arvensis* species, and not to the *Myosotis palustris*.

But why *Myosotis*? This Greek compound surely means "mouse-ear," and what have these plants to do with the auricular organs of mice? Why, their leaves were supposed to resemble in form the ear of *Mus domesticus*. The name of "scorpion-grass" originated in the fact that the top of the stem coils round while the buds are unblown, like a scorpion's tail. It is strange how quick the common people have been to detect these analogies, and to perpetuate them in the appellations they have bestowed on the flowers of the meadow, the wood, and the green lane.

The singularly beautiful name of the *Myosotis palustris*— we mean its common and non-scientific name,—is ascribed, in a well-known German legend, to the dying knight who, having ventured at a dangerous spot to pluck a handful of the bright blue blossoms for his lady-love, fell into the stream, and as he sank, flung the dear-bought spoil towards her, exclaiming, "Forget me not!"

A more probable origin is suggested by Miss Strickland. "Henry of Lancaster (Henry IV.)," she says, "appears to have been the person who gave it its emblematical and poetical meaning, by uniting it, at the period of his exile, with the initial letters of his watchword, *Souveigne vous de moi;* thus rendering it the symbol of remembrance, and, like the subsequent fatal roses of York and Lancaster and Stuart, the lily of Bourbon, and the violet of Napoleon, an historical flower."

We have said that the scorpion-grass belongs to the natural family Boraginaceæ, which receives its name from the common borage, a bright blue flower with very rough leaves. All its members are rough or hairy, except those which, like the forget-me-not, become smooth from living partly under water. The black stalks of the borage burn, it is said, like match-paper, and its root enters largely into the composition of rouge. Its flowers were at one time held in great respect as a wholesome bitter ingredient for a tankard of ale. According to Pliny, "if the leaves and flowers of the borage be immersed in wine, and that wine drunken, the potion will make men blithe and merry, and drive away all heavy sadness and dull melancholy."

Burton, in his "Anatomy of Melancholy," also says of it—

> "Borage and hellebore fill two scenes,
> Sovereign plants to purge the veins
> Of melancholy, and cheer the heart
> Of those black fumes which make it smart."

Most of the Boraginaceæ are weeds, but they include a few ornamental garden-flowers; as, for example, the Peruvian heliotrope—the "cherry-pie" of the children—which is well known for the fragrance of its blue blossoms. Its Greek name refers to an ancient belief that it always "turned" to meet the sun; but neither heliotrope nor sunflower exhibits any such devotedness towards the great "orb of day." The poet's comparison—

> "As the sunflower turns to his God, when he sets,
> The same look that he turned when he rose"—

is very pretty and suggestive, but unfortunately it is not true.

The Lilies.

Are we justified in classing these among our summer flowers? Well, the lily of the valley may, perhaps, be more justly claimed by Spring, as it generally unveils its beauty in the month of May; but the water-lily belongs to Summer; and, at all events, it will be most convenient to speak of them in this category.

The lily of the valley (*Convallaria majalis*),—the May-lily of old writers,—has long been a favourite type of retiring modesty and tender loveliness. It affects the silence and solitude of the woodlands, where, in the shadow of broad leaves and sweeping branches, the inquiring botanist discovers—

> "Like detected light,
> Its little green-tipt lamps of white."

Shakspeare, who neglected nothing, refers to its gentle humility of attitude :—

> " Shipwrecked upon a kingdom where no pity,
> No friends, no hope, no kindred weep for me;
> Almost no grave allowed me ! like the lily
> That once was mistress of the field, and flourished,
> I 'll hang my head and perish."

Our lily is a native of cold and temperate countries, and never shakes its pendant bells at the bidding of a hot Eastern breeze. It is very abundant in Norway. That agreeable writer and observant traveller, Henry Inglis, says :—" It stood everywhere around, scenting the air, and in such profusion, that it was scarcely possible to step without bruising its tender stalks and blossoms. I have not seen this flower mentioned in any enumeration of Norwegian plants, but it grows in all the western parts of Norway in latitude 59° and 60°, wherever the ground is free from forest, in greater abundance than any other wild-flower."

As it will not live in hot countries, it cannot be the "lily of the field" which furnished our Saviour with so fruitful a text for warning and instruction. This, in all probability, was the yellow amaryllis, or *Amaryllis lutea*, a flower bearing some resemblance to our yellow crocus, but much larger, and with broader leaves. Its delicate blossoms escape from an undivided spathe, or sheath, and are bell-shaped, with six clefts and six stamens, which are alternately short and long. The flower seldom rises more than three or four inches above the soil, accompanied by green leaves, which, after the flowering has passed, continue to preserve their freshness throughout the winter.

But some authorities are not content with the yellow amaryllis, and put forth as the *true* " lily of the field," either the narcissus, or the golden lily, or the stately crinum, according to their several tastes.

Not connected with these flowers by any botanical relationship, and surpassing them all in beauty, is the *Water-lily* (*Nymphæa alba*), whose large round leaves and full white blossoms are the glory of so many of our secluded lakes and quiet streams. Everybody knows old Izaak Walton's quaint eulogium on the strawberry: " Doubtless God *could* have made a better fruit, but doubtless He never hath." In like manner I am inclined to say: " Doubtless God might have created a fairer flower, but doubtless He never hath." Alas! like most things rare and beautiful, its existence is very brief! pluck it, and straightway it vanishes,—like a poet's dream, the moment he attempts to realise it.

It is sometimes asserted of our wild water-lily that it retires below the surface of the stream shortly after noon, remaining in the liquid depths during night, and rising again into the light of day at early dawn. Those who are acquainted with the haunts and habits of these beautiful flowers know that this is not strictly correct, as they may often be seen, " by the pale moonlight," lying folded above the water. It is not impossible, however, that *some* may sink ; and certain it is, that as the sun sets they close their silver vases.

> " Broad-leaved are they, and their white canopies
> Are upward turned to catch the heaven's dew."

So says Keats ; but this is true only while the sun is asserting his supremacy in the azure sky. And then, the spectacle of a calm, rush-fringed pool, nestling in the shadow of some ancient elms or drooping willows, and brightened by the

FIG. 54.—" Brightened by the uplifted cups of our delicate naiads."

uplifted cups of our delicate naiads, is a scene of surpassing beauty. We turn from this favourite flower regretfully, "murmuring," as novelists say, Mrs Hemans's graceful apostrophe :—

> " Oh ! beautiful thou art,
> Thou sculpture-like and stately river queen,
> Crowning the depths as with the light serene
> Of a pure heart !

> "Bright lily of the wave!
> Rising in fearless grace with every swell,
> Thou seem'st as if a spirit meekly brave
> Dwelt in thy cell!"

Permit me, reader, another quotation. I take it from your and my favourite, Wordsworth :—

> " Rapturously we gather flowery spoils
> From land and water; lilies of each hue,
> *Golden and white*, that float upon the waves,
> And court the wind."

The lily of golden hue, the *yellow water-lily*, is the *Nuphar lutea* of botanists. The country people, on account of its peculiar scent, most unpoetically call it "brandy-bottle." It is far more plentiful than the regal nymphæa; its flower is not so full of petals; and it is by no means so handsome. Yet, with its smooth, glossy leaves, and golden cups, and long floating stems, it favourably attracts the eye. We are told that "its roots are nutritious, and are frequently powdered and eaten for bread in Sweden;" that, mixed with the bark of the Scotch fir, they form a cake much relished by the Swedes; in which case the Swedish palate certainly cannot be censured as fastidious. "These roots are also burnt on the hearths of farmhouses, because their smoke is reputed to drive away the crickets, whose chirping is sometimes too loud and shrill to be deemed musical." Assuredly this is untrue of many parts of England, as the cricket is popularly supposed to be "lucky," and no old country-wife would allow it to be driven away from her sanctum.

The water-lily of the East, the beautiful Lotus,—the *Nelumbium speciosum*, which is figured on so many Egyptian and Indian monuments,—is rich in blue and red, as well as in white blossoms. These are said to sink quite below the surface in the evening and during the night shadows; whence Moore says of them, with his artificial prettiness—

> "Those virgin lilies, all the night
> Bathing their beauties in the lake,
> That they may rise more fresh and bright
> When their beloved sun's awake."

It was formerly abundant on the Nile, and the Egyptians consecrated it to their supreme god, Osiris; but, with the splendour and mysticism of ancient Egypt, it has completely passed away.

But it is scarcely less prized by the Hindus, who have also consecrated it to one of their deities. A traveller thus speaks of the sacred Ganges in connexion with it:—"The rich and luxuriant clusters of the lotus float in quick succession upon the silvery current. Nor is it the sacred lotus alone which embellishes the wavelets of the Ganges; large white, yellow, and scarlet flowers pay an equal tribute; and the prows of the numerous native vessels navigating the stream are garlanded by long wreaths of the most brilliant daughters of the parterre. India may be called a paradise of flowers: the most beautiful lilies grow spontaneously upon the sandy shores of the rivers, and from every projecting cliff some shrub dips its flowers in the waters below."

No reader of English poetry but is familiar with Tennyson's "Lotus-Eaters"—a poem founded on the old myth of a people who lived upon the insane root that takes the reason prisoner, and, beguiled by its sweet intoxication, abandoned themselves to a state of dreamy repose.

> "And round about the keel with faces pale,
> Dark faces, pale against that rosy flame,
> The mild-eyed, melancholy Lotus-eaters came.

> "Branches they bore of that enchanted stem,
> Laden with flower and fruit, whereof they gave
> To each, but whoso did receive of them,
> And taste, to him the gushing of the wave
> Far, far away did seem to mourn and rave
> On alien shores; and if his fellow spake,
> His voice was thin, as voices from the grave;
> And deep asleep he seemed, yet all awake,
> And music in his ears his beating heart did make.

> "They sat them down upon the yellow sand,
> Between the sun and moon upon the shore;
> And sweet it was to dream of Fatherland,
> Of wife, and child, and slave; but evermore
> Most weary seemed the sea, weary the oar,
> Weary the wandering fields of barren foam.
> Then some one said, 'We will return no more;'
> And all at once they sang, 'Our island home
> Is far beyond the wave; we will no longer roam.'"

But the lotus of poetry is not the *Nelumbium speciosum*. There is some difficulty in identifying it with any modern plant; but the general opinion seems to be, that it was the *Zizyphus lotus*, a species allied to the *Zizyphus jujuba*, and included in the Buckthorn family (*Rhamnaceæ*).

The reader will be by this time aware that of the plants called by the general English name of "lily," some have very little kinship to each other, and others none at all. The little garden flowers named *Lilium* (from the Celtic word *lis*, "whiteness") are mostly very handsome. Ben Jonson, speaking of the ordinary lily, says, very finely—

> "It is not growing like a tree
> In bulk, doth make man better be,
> Or standing long, an oak, three hundred year,
> To fall a log at last, dry, bald, and sear:
> A lily of a day
> Is fairer far in May,
> Although it fall and die that night,—
> It was the plant and flower of light."

The white garden-lily is a native of the Levant, but has become thoroughly naturalised in England, and is one of the commonest but most admired ornaments of our cottage-gardens. The old herbalists thought highly of its medicinal properties, and pronounced it a certain remedy for the bite of a serpent. It is true, at all events, that its bruised petals are an excellent cure for any ordinary wound or bruise.

Our ancestors, among their other superstitious fancies, entertained the extraordinary belief that the price of a bushel of wheat in the ensuing season was foretold by the number of white cups which crowned the white lily's stem, each cup being estimated at one shilling. I opine that our modern farmers would feel dissatisfied if the Mark Lane averages were regulated by this simple standard.

The common Turk's-cap lily (*Lilium martagon*) is identified with the ancient hyacinth, the "sanguine flower inscribed with woe." The orange lily (*Lilium bulbiferum*) is a native of Southern Europe. When the Dutch were at feud with the House of Orange, they were accustomed to root up this

FIG. 55.—Lily of the Valley.

flower from their gardens, as some solace to their indignant feelings.

The garden lilies belong to the natural family of the *Liliaceæ*, which includes the following sub-orders:—

1. *Tulipeæ*, tulip tribe; bulbous plants, with the segments of the perianth scarcely adherent in a tube.

2. *Hemerocallideæ*, day-lily tribe; bulbous plants, with a tubular perianth.

3. *Scilleæ*, or *Alliæ*, squill or onion tribe; bulbous, with black and brittle testa.

4. *Anthericeæ* or *Asphodeleæ*, asphodel tribe; roots fascicled or fibrous, leaves neither coriaceous nor permanent.

5. *Convollarieæ*, lily of the valley tribe; stem developed as a rhizome or tuber.

6. *Asparageæ*, asparagus tribe; stem usually fully developed, arborescent, branched in some cases, and leaves frequently permanent and coriaceous.

7. *Alonieæ*, aloes tribe; stem usually developed, arborescent, with succulent leaves.

8. *Aphyllantheæ*, grass-tree tribe; characterised by a rush-like habit and membranous imbricated bracts.

THE GENTIANS.

Let me now direct your attention, reader, to a pretty plant, of very elegant appearance: crowned, as it is, by a cluster of rosy flowers, it would not disgrace our well-kept parterres. It is called the common Centaury (*Erythræa centaurium*). You will never see it in the fields side by side with the *Delphinium*; but in July and August will meet with it frequently on the borders of woodland paths and open glades.

Would you create for yourself by the study of nature a source of enjoyment equally pure and inexhaustible, adopt a method of classification for your own use, and, to facilitate you in the task, take for your types those plants which are at once the commonest and most characteristic of each season. Quite at your ease, you may begin your analysis by examining the parts which, like the calyx and the corolla, most attract your attention. The most rational plan, however, would be, to commence with the seed, and to follow it through all the wonderful phases of its life, from the development of the embryo to the maturity of the fruit. Unfortunately, we are all compelled to take time into account; time is so much more precious than money,—it is the measurement of our existence. Undoubtedly, the mind, with its gigantic strides, like those of an Homeric god, tends to overleap the confines both of time and space. But the senses, without whose co-operation the intellect could not create science, never fail to remind us that we are, alas! but mortals. By this incessant appeal to order, we are under the necessity of doing, not what we *would*, but what we *can*. And the part we really play is, consequently, much more modest than that which we love to imagine ourselves as playing.

But to return to our flowers.

What see you in the little centaury which you hold in your hand? (Fig. 56.)

In the first place, a corolla with five petals of a delicate rose-hue, very pleasant to the sight.

Take care! those foliola are not *petals*, if you give that name to the free parts of the corolla. Look at them thoroughly. Your foliola are prolonged at their base in a narrow tube, which is easily removed. If you had begun here,—if, instead of proceeding from the top to the bottom, you had, in your analysis, proceeded from the bottom to the top, you would have acquired a wholly different view of things. You would

FIG. 57.—The Common Centaury.

have said that the corolla is tubular, greenish, with a rosy limb, deeply divided into five lobes. And in so doing, you would have run no risk of deceiving yourself. The indications given by Nature herself are the most precious; they are the lessons of a teacher who cannot err: never pass them by with indifference or neglect. In your study of the different parts of a vegetable, follow, as far as possible, the actual movement of the sap.

The calyx of our gentian has, like its corolla, the form of a five-divided tube; which, indeed, is one of the usual characters of the Gentian family.

But it is important here to take notice of this fact, because it is not, as at first sight you would suppose, the corolla, but the calyx, which encircles the base of the ovary. The tube of the corolla stops towards the middle of the latter organ, and nearly on a level with the linear divisions of the calyx. You must be careful not to confound with these calycine divisions the green foliola which lie around the base of the flower, and which are neither more nor less than abortive leaves.

Now call to mind that the flower is an union of concentric whorls, or of rings set one within another. The *staminal* whorl and the *carpellary* whorl, surrounded by a double perianth (corolla and calyx), are here composed—the first, of five stamens, and the second of a bi-lobed ovary, surmounted by a twisted style. We may now examine more closely the reproductive organs.

The stamens are inserted upon the top of the tube of the corolla, and if you look at the base of their filaments you would be inclined to pronounce it a foliaceous expansion, or, rather, a metamorphic doubling of the corolla. Suppose the stamens to be the result of the transformation of the petals, the filament would be the "claw," and the anther the "limb" of a foliole. At least, theory would tell you so. But observation will show that these are not petals changing into stamens, but, on the contrary, stamens changing into petals; as is seen in the sterile (or "double") flowers of many of our ornamental plants, and even of some of our fruit trees. How, then, shall we conciliate theory with observation? Look, and you shall find.

Observe the anthers which surmount the filaments. There is something peculiarly characteristic in them. As they open and spread abroad the pollen, they visibly coil themselves up in the form of a spiral. (Fig. 58.) Owing to this twisting, they are found more or less inclined upon their filaments, and are gifted with a considerable mobility. Thoroughly to understand the relation between the continuous anther and the filament, we must examine the stamens before their expansion, while they are still folded up within the floral bud, their matrix. The anthers are then quite straight, and, with a magnifying-glass, you can easily see how they are inserted, by the lower part of their back, upon the top of the filament, whose (so-called) *connective* prolongation separates the two lobes of the pollen receptacle. The anthers are then *introrse* (*introrsum,* inwardly)—that is to say, their face being inclined inwards, they look towards the centre of the flower, occupied by the style, a filiform prolongation of the ovary; the apex of the style (stigma) is thick, globulose, and of a glandular structure. The fruit, resulting from the metamorphosis of the ovary, is an elongated fusiform capsule, composed of two lobes, each containing a very large number of extremely small seeds.

FIG. 35.—Anthers of the Centaury.

The characters we have just enumerated apply, or the majority of them, to the interesting family of the Gentianaceæ —a natural group of plants, nearly all remarkable for their bitter, febrifugal, and anti-scrofulous properties. The flowering cymes of the common centaury are very frequently employed as a substitute for the medicinal gentian, so well known as a valuable tonic.

The medicinal gentian is the *Gentiana lutea,*—a plant growing about three feet high,—which thrives abundantly on the Pyrenees, and the Alps of Switzerland and Austria, at an elevation of 3000 to 5000 feet. It is not, however, so common now as formerly on the Alpine heights, owing probably to its great consumption, but it is spreading into many districts of Central Europe.

Frequent enough in the vicinity of Paris is the *Chlora perforata,* a gentian remarkable for its glaucous leaves and yellow terminal flowers.

The *Gentiana kurroo* of the Himalayas, and the British species, *Gentiana campestris* and *Gentiana amarella*, possess the tonic properties of the family. The *Cheritta* of the pharmacopœia is the herb and root of *Agathotes chirayta* (*Ophelia chirata*), a herbaceous plant which flourishes in the Himalayas.

We must not omit a reference to a Lilliputian gentian, the *filiform gentian* of Linnæus, and the *Exacum filiforme* or *Cicendia filiformis* of other botanical authorities.

Its stem, from two and a half to four inches in length, is embellished with radical oblong leaves, disposed in fours, and short caulinary leaves, opposite and linear; the corolla is yellow, the calyx has four triangular lobes; the stamens also are four. It is this predominant number which has induced some botanists to elevate our little gentian into a species of *Exacum* or *Cicendia*,—two genera, of which the first was named by Adamson, the second by De Candolle. As for the tiny *Gentiana pusilla*, or *Exacum pusillum*, we may look upon it as a simple variety of the *Exacum filiforme*, differing from the latter only in its shorter and feebler stem, in the somewhat narrower divisions of its calyx, and the tint of its corolla, which is of a paler yellow, sometimes inclining to rose.

Botanists, or lovers of flowers, may grow as passionately fond of gentians as some persons do of tulips or hyacinths. But it is not in England or Scotland, it is in the Alpine pastures of Switzerland only, that you can hope to satisfy your *Gentianomania*.

An Alpine Excursion.

Permit me, dear reader, to set down a few hints, in case you should at any time be disposed to make a pilgrimage into the Golden Land, or El Dorado, of botanists and geologists.

Before you plunge into the Alps, you will meet, in the sub-

FIG. 59.—An Alpine Landscape.

alpine regions, among the valleys which intersect and the meadows which clothe the lower spurs of the Jura, the *Gentiana campestris*. It is a plant of from five to six inches in height, whose blue, five-lobed corolla, with its velvety gorge, changes into yellow when dried; the two outer teeth of the calyx are elliptical, and much larger than the others. Your attention will hardly be drawn to this tiny gentian among the crowd of more beautiful and attractive plants blooming around it.

One excursion which you should not fail to make is the ascent of the Dent de Jáman,—the hieroglyphic summit of one of those charming mountains which mirror themselves in the Lake of Geneva. This classical, and, moreover, very easy ascent, has the advantage of carrying you up a series of terraces, so that the character of the vegetation changes rapidly. The acclivity begins at the little town of Montreux, situated near the point where the blue arrowy Rhone pours its waters into the enchanted lake. From Montreux to the village of Glion, you will be delighted to greet your old acquaintances: the familiar faces of your native fields, meadows, and woodlands. But soon a difference in the character of the flora forces itself upon you. Species which are rarer at home become tolerably common, as, for instance, the yellow digitalis (*Digitalis lutea*), so easily recognised by its cymes of tiny flowers. The vulneraria (*Anthyllis vulneraria*) is as widely diffused as the trefoil in our English pastures.

A little above Glion,—a picturesque village apparently hung over the lake—the *sub-alpine region* commences. Around villas and mansions, nearly all inhabited by English families, you will find in abundance the English mercury (*Chenopodium Bonus Henricus*), and in the shady hedges the narcotic Herb-Paris (*Paris quadrifolia*).

The apparition of the *Astrantia major*, which resembles an artificial or fancy-created flower, warns us that we are passing beyond the limits of the ordinary flora. I believe that we have no representative in England of that singular umbellifer. The Alpine pastures of the narrow-ridged Mount Caü, which resembles the back of a dromedary, exhale a fragrance like that of the famous Swiss tea, so much extolled as a remedy against cholera. The odour of the hay-lofts attached to the châlets—true shepherd's huts—which rise at intervals along the back of Mont Dromedary, is so penetrating as to produce headache. The hay owes its aromatic fragrance to the musk-chervil (*Myrrhis odorata*), whose strong stems form such thick luxuriant pasturages; to various orchideæ, particularly to the *Nigritella suaveolens*, remarkable for the intense colour—nearly black—of its flowers; and finally, to the gentians, whose scent is strongly brought out by drying.

The rich close sward which borders on the Dent de Jaman provides the herboriser with more than one agreeable surprise. You will be struck by the beauty of the flowery tufts of the *Linaria Alpina*, rejoicing in a deep sapphire blue. You will also have an opportunity of making acquaintance with a campanula which is abundant on Mont Cenis (*Campanula Cænisia*); its beautiful terminal flower, of a pale blue, is characterised by the long hairs which line the opening of the corolla.

Among the gentians, those great ornaments of the Alpine pasturages, we shall direct the attention of our readers to—

The purple gentian (*Gentiana purpurea*), and the spotted gentian (*Gentiana punctata*). These are distinguished by their plentiful appearance: their large oval leaves, and the height of their vigorous stems, recall those of the yellow gentian. The features which separate these two species are not very distinct: the corolla of the former is purple without and yellow within; that of the latter is of a bright yellow, marked by spots of deep purple, which, however, are not permanent; the calyx is campanulated, with upright and lanceolate foliola.

The *Gentiana acaulis* contrasts singularly with the preceding species, its stem being so short that one is almost tempted to deny the existence of any; its large corollas, of a bright celestial blue, lie on the ground as if they had fallen fresh from a bouquet. We must not confound this species with the *Gentiana pumila*, a much smaller plant, with a very elongated calyx, which grows abundantly on the turf of Mont St Bernard.

The *Gentiana verna* and the *Gentiana nivalis*, with a corolla of the finest azure, inhabit the loftiest points of the Alps, where all vegetation begins to disappear. The former, or the gentian of spring, flowers, in these frozen regions, in June and July; it is one of twenty-four phanerogamous plants of the last vegetable station of Mont Blanc. This station is formed by a series of vertical layers of protogene, which separates the upper part of the Glacier des Bossons from that of Taconay. The débris of the rock, decomposed under the influence of atmospheric agencies, form, in the midst of the *nevé*, tiny flowering parterres—oases in the desert, islands in the vast ocean of ice and snow. There, sheltered by the rocks, and warmed by the sun, and refreshed by the snow, which rapidly melts in summer, these pretty plants thrive and grow beautiful, though their brief existence is summed up in a few short weeks.

According to Charles Martins, the phanerogamous plants which flourish at an elevation of 10,000 feet are the following :—

Mean temperature, 47° to 36°.

1. Gentiana verna.
2. Silena acaulis.
3. Draba frigida.*
4. Draba fladnizensis.
5. Cardamine bellidifolia.
6. Cardamine resedifolia.
7. Potentilla frigida.
8. Phyteuma hemisphericum.
9. Pyrethrum Alpinum.
10. Erigeron uniflorum.
11. Saxifraga bryoides.
12. Saxifraga Groenlandica.
13. Saxifraga muscoides.
14. Saxifraga oppositifolia.
15. Androsace Helvetica.
16. Androsace pubescens.
17. Lazula spicata.
18. Festuca Halleri.
19. Poa laxa. †
20. Poa cæsia.
21. Poa Alpina.
22. Trisetum subspicatum.
23. Agrostis rupestris.
24. Carex nigra (Cariceæ).

THE PIMPERNEL.

Accompany me to the corn-field; not for any discourse upon the state and prospect of the crops, or on the comparative value to man of wheat or barley, but for the sake of the

---

* *Draba aretoides* has been found on Chimborazo at 16,000 feet.
† *Poa annua* grows at an elevation of 7400 feet.

little red flower which shines like a star among the growing harvest.

You cannot mistake it, for, with the exception of the tiny chaff-weed, the smallest wild plant which bears a distinct flower, it is the only scarlet blossom in the wheat-field, except, indeed, the red poppy, which every good farmer seeks to banish from his land. Mark me,—I say the only *scarlet* flower; for there are several—as, for instance, the pheasant's eye, or Adonis—of a deep crimson.

The pimpernel belongs to the Primrose family, or Primulaceæ. It has a five-cleft calyx, and a monopetalous corolla. Its stamens, of an equal number, are inserted on the corolla, opposite its segments. It is a *meteoric* flower; so-called, because it keeps itself shut during wet or cloudy weather. Hence, it is known among country people as "the shepherd's warning" or "poor man's weather-glass." And Darwin, enumerating the various signs of rain, says of it—

> "Closed is the pink-eyed pimpernel;
> In fiery red the sun doth rise,
> Then wades through clouds to mount the skies:
> 'Twill surely rain, we see't with sorrow;
> No working in the fields to-morrow."

It should be added, however, that if the rain continue for several days, the pimpernel will lose its sensibility, and cease to act as a natural weather-glass.

And here we may observe, that singular as is the habit of the flowers anticipating rain by folding their petals within their calices, the way which the Siberian sow-thistle has with it is still more curious. This plant, during that clear weather which most flowers affect, keeps entirely shut; but as soon as a thick mist overspreads the earth, or a cloud obscures the bright face of heaven, it begins to open its light blue corolla.

Everybody knows, or should know, that when the robin looks sad and drooping, and ceases to greet you with his wonted blithesome strain, "foul weather" is at hand. Many animals, by their peculiar habits, afford equally certain indications of approaching atmospheric changes.

This does not seem strange to us; we account for it by the instinct which every animal possesses, in a larger or smaller degree. But the same anticipatory faculty is possessed by several plants; they feel the increasing moisture of the air long before it can be detected by ourselves. Thus, when a storm is at hand, some species of anemones fold up their blossoms; the fragrant flowers of the wild pink convolvulus wind themselves together; the awns of the wild oat, and the sweet-scented meadow-grass, stand in an erect position, and the clover leaves are drawn closely up.

Naturalists, says Pratt, are unable to discover why some plants should be affected by moisture and others not; but the regular changes of these natural barometers seem a providential arrangement to supply certain wants of the flowers in which they occur. We may draw this inference from the different positions of several flowers according to their circumstances. Thus the poppy, when in bud, hangs down on its stem, and preserves its petals from rain and wind; but as soon as it is fully developed, and has acquired strength, and the sun's rays are necessary to perfect its colours, it expands to the full light of day. The violet, again, while its seed is forming, shades the capsule by its purple corolla; but as soon as the seeds are ripe, and they are required to spring to some distance from their capsules, the flower immediately rises up with the cup for its support, and flings abroad its offering on the earth's maternal bosom. Adaptations of this kind are frequent and striking in the vegetable kingdom, and surely one is justified in regarding them as the work of an all-powerful and all-wise Creative Mind. Look, for instance, at the orchis: it grows on the ground in Europe, and is consequently provided with roots formed of large lobes; but when it festoons the pillar of the virgin forests of the New World, its roots are formed of a number of fibres, so that they may penetrate the bark of the tree.

But to return to our pimpernel. It was at one time called *Centunculus*, from *cento*, a covering, because it spread in such abundance over the cultivated fields. Its botanical name was afterwards changed to *Anagallis arvensis*. *Anagallis* signifies "to laugh," and there existed an old belief that a decoction of the pimpernel acted as a remedy against melancholy, and a provocative of mirth.

The seeds of this plant are very numerous. They are enclosed in small capsules, and eaten by the birds.

There is only one other *British* species of pimpernel, the *Anagallis linetta*, or bog pimpernel, which it would be unpardonable if the botanist omitted to notice, so delicately beautiful are its pale-rose blossoms and tiny clusters of leaves. As its name indicates, it is found only in marshy localities.

The blue pimpernel (*Anagallis cerulea*), though not a native of England, is found occasionally. It is described as growing in beautiful little tufts about the hills of Madeira, and enlivening them by its cheerful colour, which may bear comparison with the azure of the sky.

Fig. 60.—"The Wheat-field, with its mass of emerald waves."

And here we will take leave of the wheat-field, with its mass of emerald waves, now beginning to wear a golden glory on their crests, as they ripple in the genial sunshine.

Animals.

The Mole, the Staphylinus, and the Mole-Cricket.

Do you hear that noise? It seems to issue from beneath yonder heap of pebbles at the foot of the garden-wall. Surely the stones are moving; they seem to be walking alone, and of their own impulse, for I cannot see anything to set them in motion.

Let us draw near and examine the mystery. "Ah, what a hideous black creature!" It is retreating in a terrible state of alarm, as if it felt itself pursued by some formidable enemy.

This "hideous creature" is known to French gardeners by the name of *courtilière;* to naturalists, as a species of *Staphylinus.* (Fig. 61.) Its great and persistent adversary is the mole: a mammal at war with an insect!

FIG. 61.—The *Staphylinus olens.*

Watch well, I pray you, the mole's movements, which you can do the more easily that here, contrary to his ordinary custom, he is wandering in the open day; the light blinds him,—accustomed as he is to pass his life in the subterranean galleries which he excavates by his own labour. But if he does not see us, he hears us; the sound of our footsteps was sufficient to make him prick up his ears (if we may so speak of a mole), and he remains motionless. Do not stir, or he will betake himself to flight, and we shall lose an excellent opportunity of being present at a very curious spectacle.

He is now reassured. He recommences his manœuvres, pushing before him every little pebble which he meets with. For this purpose he employs his elongated snout, exactly as a pig grubs among the uncleanness of his sty. But his next movements are not of a porcine character. With feet broad as battledores, the mole, while manœuvring with his nose, quietly pushes aside every clod which threatens to obstruct his progress. These sidelong, abrupt, and jerking movements remind you of those of a dog, seeking with his paws to enlarge the opening of the burrow wherein a rabbit has taken refuge. The mole has thus the habit of a hunting-dog; and, to complete the resemblance, he stops at intervals in his scratching, and shakes the dust off his head. One is quite surprised to see a little mammal executing the movements we are wont to regard as peculiar to an animal much larger than he is.

The beetle quits in affright the heap of stones where it had hoped to find an asylum; it now crosses our path, holding itself erect, with a menacing air, and its tail armed with a forked barb. The mole follows in close pursuit: who would have believed he could run so quickly? Let us bar his passage, and study him at our leisure.

The first thing we remark is his glossy hair, which is softer to the touch than the finest velvet. Where are his eyes? Blow aside the hair which covers his face. There they are, and they resemble miniature pearls of a shining black.

How could Aristotle say that the mole had no eyes? To

believe it you must read the assertion for yourself. And here are the very words of the authority who, for so many centuries, was accepted as infallible :—

"All viviparous animals have eyes, *except the mole*" (πλὴν ἀσπάλαχος).

Then, as if a sudden doubt had seized him, and he were frightened at his own statement, the illustrious Stagyrite hastens to add: "We might, perhaps, strictly admit that he has." But another change comes over the spirit of the philosopher's dream; his hesitation vanishes, and he immediately repeats and justifies his former assertion in these terms :—

"But, carefully considered, the mole does *not* see, because he has no apparent eyes externally" (ὅλως μὲν γὰρ οὔθ' ὁρᾷ, οὔτ' ἔχει εἰς τὸ φανερὸν δήλους ὀφθαλμούς).*

These last words denote—I beg pardon of the manes of a great philosopher—an absolute want of observation. Evidently, Aristotle had not taken the trouble to *look* before he made his statement. And do not think that this curious indifference was peculiar to the great master of the Peripatetic School; it characterises more or less all the philosophers of antiquity, as well as too many who have followed in their footsteps.

Pliny has simply translated Aristotle when he says :— "Among quadrupeds the moles are wanting in the sense of sight" (*quadrupedum talpis visus non est*).†

---

* Aristotle, "Hist. Animal.," i. 9.
† Pliny, "Hist. Nat.," xi. 52.

But it is a curious thing that both Aristotle and Pliny maintain, that if you lift up the skin where the eyes *ought* to be, you will perceive the organs of vision. How could they remove the skin without distinguishing in it the eyes, like black and brilliant beads? Did they practise anatomy, like their own imaginary mole, without eyes? The whole matter would be inexplicable if we did not take into account that force of inertia which binds man in chains of iron, in the moral world as well as in the physical.

To see, to observe; to retrace one's steps, that one may see and observe more distinctly; is a labour repugnant to the human mind. To create systems, in order that he may proclaim himself a great doctrinal teacher, is the work which flatters man, or the creative power of his imagination. Centuries of effort are needed before he can disentangle himself from his self-woven thrall in presence of the phenomena of nature.

A striking peculiarity in the mole's structure is his hands, or feet, with their five fingers, or toes, turned outward, and their curious resemblance to the human hand. Few animals exhibit a similar conformation. Everything in the structure of the fore-limbs indicates the animal's burrowing instinct,—the length of the bone, which corresponds to the human *radius*, or fore-arm,—the breadth of the hands,—and the bend of the arms, which are so fashioned that the elbows project outwardly.

Does the mole's burrowing instinct lead it in quest of insects or of vegetable roots?

According to the old traditional belief, the mole feeds upon roots. From time immemorial, it has been looked upon as an animal so destructive, that, in every country, its destruction has been encouraged by large rewards. Well, this belief, transmitted from generation to generation, owes, like so many other traditions, its authority to its antiquity, and is devoid of foundation.

The mole is an essentially carnivorous animal, and no more lives upon roots than the dog or the cat. He is pre-eminently the hunter of the white-worm and beetle; and therefore, instead of vowing its extermination, we ought to take every possible means to preserve and multiply his race. These absurd traditions and credulous notions, wholly without any experimental confirmation, frequently lead us to take steps in diametrical opposition to our own interests.

Moles are particularly partial to meadows which are somewhat damp, as, for instance, those where the leafless colchicum displays, in autumn, its pale-rose flowers. In summer, the fields are covered with mole-hills. It will be said, perhaps, that these conical heaps of earth are injurious to vegetation. But that will be an error, contradicted by observation. Meadows besprinkled with mole-hills grow excellent hay, if care be taken to level them; for the earth thus distributed serves as manure. If they are visited by the moles, it is because these animals find there a plentiful supply of the *rhizophagous* (root-eating) insects, on which they feed.

The forest is also a favourite haunt of the moles. Apparently they find, under the layers of leaves and roots, so rich in larvæ of every kind, the wherewithal to satisfy amply their insectivorous tastes. It is the mole which generally produces that rustling of the dry leaves the wanderer is so apt to attribute to a snake or an adder. Stand still for a moment, and patiently watch. Do you see that undulatory movement? Thrust your stick rapidly into the uplifted heap. There is our persevering hunter; he struggles hard to escape from his terrible enemy, but, with a little alacrity, you will not find it difficult to capture him.

Moles are among the most prolific of mammals; and, in fact, were it not so, their race would have been long ago exterminated. We may, perhaps, venture to say, that by multiplying so prodigiously, they wish to do us a service in spite of ourselves. How tender is the solicitude of nature for the ungrateful human species!

To see the marvellous qualities ascribed to the mole by the ancients, one would suppose that they had made him the object of their special study. Yet, as we have shown, they could never have watched his habits with any degree of patience. They saw everything through the delusive prism of their imagination. As a proof, we will tell you what they said of the mole.

"Since this animal has been doomed to a perpetual blindness, and lives interred beneath the surface of the earth, like the dead, he possesses, by way of compensation, some extraordinary qualities. His subterranean existence renders him, of all animals, the most capable of religion (*nullum religionis capacius animal*). To acquire the gift of second-sight, you must eat the heart of a mole, while still beating, and freshly plucked from the animal's body. To cure toothache, suspend to your neck the tooth of a live mole. Lymphatic people will gain in strength if sprinkled with a mole's blood. The ashes of a mole are a sovereign remedy for scrofula; some recommend for this disease the animal's liver, others the right foot, and others the head. The earth of mole-hills, fashioned into pastilles, and preserved in a tin box, is an excellent cure for all kinds of tumours, and especially for abscesses on the neck."*

Such, according to Pliny's report, are the virtues of the mole, as taught by the Magi. The Middle Ages adopted this teaching, and even to-day, in obscure rural districts, you will meet with superstitious notions which remind you of the ideas of the ancient wise men and necromancers of antiquity.

We have thus summarised the natural history of our *hunter*, let us now say a few words respecting the *game* he pursues.

The insect before us is the *Staphylinus olens*. Its study has been much neglected, probably on account of its repulsive appearance. But, conquering our repugnance, let us take the creature between our forefinger and thumb. See how vigorously it defends itself! Its forked appendage is not formidable, it is too soft; but take care of its mandibles! With these hard, horny, pointed pincers, it pricks the skin and draws blood. Now, bring your nose close to the frightful black insect at the very moment when it appears the most

* Pliny, "Hist. Nat.," xxx. 7, 12, 24.

irritated. Come! A little courage will conquer your new feeling of disgust.

What do you smell?

A pleasant odour of rennet apples! It reminds me of that diffused by another insect, much less ugly than your *Staphylinus*, the *Cicondela campestris*.

It is this peculiarity which explains the specific name of "odorous" (*olens*) given to your captive. As for its generic name, *Staphylinus*, I have no means of interpreting its etymology; for the insect's shape has no resemblance to that of a bunch or cluster,—in Greek, ςraφυλή. But this last word also signifies the *uvula*, and, perhaps, by the effort of a little imagination, the naturalist may trace a similitude between that organ of the throat and the body of the *Staphylinus*.

The *Staphylini* are characterised by a very narrow neck, which separates, as by a kind of web, the head from the thorax. In diffusing the peculiar odour of which we have just spoken, they simultaneously eject a musky volatile liquid contained in two retractile whitish bladders, situated near the anus. They run quickly, elevating their abdomen like the earwig. The antennæ, inserted in the rear of the strong mandibles, are each composed of eleven articulations, of which the first is the longest; these joints, rounded in form, are arranged like the beads of a necklace.

The Staphylini belong to that numerous section of insects whose *tarsi* are composed of five articulations, and which have thence received the name of Pentaceii

In this section they form, with some other genera, the family of *Brachelytræ*, so called because their elytra, or wing-sheaths, are much shorter than the abdomen.

Our *Staphylinus olens* is finely punctuated, somewhat hairy, and of a dull black colour. Though very common in our gardens, and wherever any putrefying substances are to be found, its habits are not very well known. For if it were generally understood that it is an essentially carnivorous animal, that it carries on a determined warfare against the caterpillar, larvæ, and especially the white-worm, far from seeking to destroy it, men would surely attempt to increase its numbers. It is a proof that the Staphylini are useful insects, that they are rare in seasons when the white-worms abound, as was the case, for example, in the years 1867 and 1868. The larva of the Staphylinus is as carnivorous as the perfect insect, which it likewise resembles in form.

To sum up: in every phase of their existence, the Staphylini render immense services to the agriculturist. It is very desirable that this fact should be generally recognised, and their rehabilitation generally proclaimed.

THE MOLE-CRICKET.

The habits of the mole-cricket are nearly the same as those of the mole. When winter approaches, it takes refuge underneath the surface of the earth, and remains benumbed and lethargic in its nest so long as the cold lasts. On the welcome return of spring, it makes its way back to the light by a vertical gallery, on which a great number of lateral galleries abut, the said lateral galleries being the roads it travels in pursuit of its prey. This subterranean work it executes with its strong forefeet, which are broad, and unguiculated, or indented, much like those of a mole. Hence its popular name of the *mole-cricket* (Fig. 62, *a*).

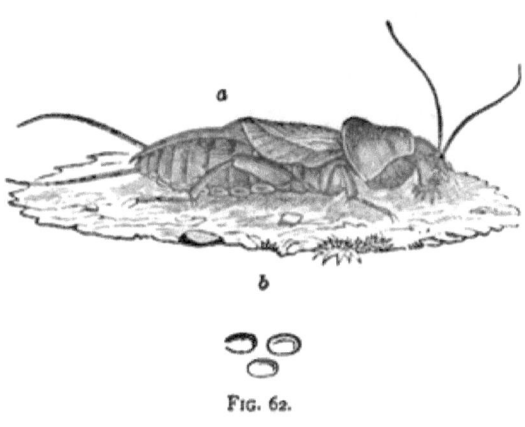

Fig. 62.

These insects (of the *Orthoptera* order) belong to the small family of the crickets—a family closely akin to that of the grasshoppers. This close kinship has been recognised by the poets, and we find them brought together in a very charming sonnet, which cannot be too frequently perused by any reader, and which may therefore be introduced as a relief to our duller prose :—

"Green little vaulter in the sunny grass,
   Catching your heart up at the feel of June,
   Sole voice that's heard amidst the lazy noon,
When even the bees lag at the summoning brass ;
And you, warm little housekeeper, who class
   With those who think the candles come too soon,
   Loving the fire, and with your tricksome tune
Nick the glad silent moments as they pass ;
Oh, sweet and tiny cousins, that belong,
   One to the fields, the other to the hearth,
Both have your sunshine ; both, though small, are strong
   At your clear hearts ; and both were sent on earth
To sing in thoughtful ears this natural song—
   Indoors and out, summer and winter, mirth."

So sings Leigh Hunt—a poet, by the way, whose heart was ever open at "the feel of June," and whose genial writings, whether prose or verse, whether delightful essays or melodious songs, should be read in the "happy summer-time," when the idler, reclining on the sunny grass, with the beauty of an English landscape around him, wants the companionship of a gentle spirit and a refined and healthy intellect.

FIG. 63.—The idler, reclining on the sunny grass.

The mole-cricket is, like the grasshopper, a child of summer. It differs, moreover, from the "cricket on the hearth" in lacking those organs of *stridulation* (excuse the word, kind reader!) which mark "the glad silent moments" with their tricksome (and sometimes inconvenient) tune. Their posterior thighs have an apparent bulging about them, but the legs are very short; so short, that our little friend could not compete with its cousin, the grasshopper, in vaulting exercises, even were it not otherwise prevented by its large abdomen. Nor is it much assisted by its wings, for though they are broad, they are not organised for rapid flight, and the mole-cricket makes but little use of them. Nature, however, has compensated it for all these disadvantages by the gift of those strong, powerful, flexible fore-feet of which I have already spoken.

The species generally met with in gardens, corn-fields, and orchards is the *Gryllotalpa vulgaris* of Latreille, identical with the *Gryllus gryllotalpa* of Linnæus. It has a brown head, garnished with rusty-coloured mandibles; the thorax is of a brownish-gray, velvety, tinged with red in the fore parts; the elytra, or wing-sheaths, which are much shorter than the abdomen, are gray, and marked by black and conspicuously prominent nerves; the wings, folded back like a fan, are about one-fourth longer than the abdomen.

The mole-cricket,—mark me!—is no more of a root-eater than the mole; it is carnivorous, like the Staphylinus. As an experiment in confirmation of this statement, we shut up one of these curious Orthopteras in a large chest filled with mould. Concealed in the galleries which it speedily constructed for its use, it fed upon larvæ, and never touched the cereals which we had sown in the earth. Here was a proof that we had to plead the cause of another of man's victims.

The mole-cricket ejects, when pursued or tormented, a blackish liquid, whose etherealised odour reminds one of the peculiar smell of certain rotten apples. The female, larger than the male, lays her eggs, which are, comparatively speaking, of a tolerable size (Fig. 62, *b*), at some depth underground. The young, when hatched, resemble their parents, except that they are white, and possess merely the rudiments of wings.

If the mole-cricket, in its subterranean progress, encounters any roots, it cuts them with its mandibles, not to feed upon them, but to get rid of an obstacle; hence the mischief of which the farmer accuses it, though this slight amount of injury is altogether outweighed by its services in destroying a swarm of insects.

Perhaps, therefore, we must not blame the farmer for the hostility with which he pursues it, especially if we are to accept as a true picture of its doings the sketch recently drawn by a popular writer:—

"It is easy to understand that an insect which undermines land in this way must cause great damage to cultivation (!). Whether the crops serve it for food or not, they are not the less destroyed by its underground burrowings. Lands infested by the mole-cricket are recognisable by the colour of the vegetation, which is yellow and withered; and the rubbish which these miners heap up at the side of the openings leading to their galleries, resembling mole-hills in miniature, betrays their presence to the farmer."

If, I say, this be a true picture, we cannot wonder at the means employed by the farmer to clear his fields of such dangerous tenants. The plan generally adopted is to dig, at intervals, a number of little trenches, which are filled up with cow-dung, well trodden down. The supposed root-eaters assemble in these warm nests; and every fourth or fifth day, a labourer, armed with a pitchfork, scoops up the manure at a single stroke, and scatters it over the ground, while another crushes the unfortunate *Gryllotalpæ* as fast as they make their appearance.

THE EARWIG.

Next to the domestic fly, the earwig is, perhaps, one of our commonest, and, let us add, one of our most troublesome, insects. Whence comes its popular appellation? From a mere fable. To amuse the silly—alas! how great their number!—marvelling, without doubt, at the spectacle of an insect's tail armed with strong pincers, some jester wished to transform it into a terrible animal; and therefore he pretended that it introduced itself into the human ear, and from thence penetrated to the brain, with the view of driving out its proprietor, —*i.e.*, the mind or spirit which animates it. Only, the originator of this absurd bugbear forgot one little fact: there is no opening by which the ear can communicate with the brain! As for the pincers, they are not so formidable as they appear. This character, however, has been considered a sufficient foundation by the naturalists, even by Linnæus himself, for the insect's scientific name, *Forficula auricularia*, which is almost literally translated by the French *oreille-pince*,—our English *earwig* or *ear-piercer*. (Fig. 64.)

FIG. 64.—The Earwig (*Forficula auricularia*).

What dress is to man, their wings are to insects; by these

we distinguish them, at the first glance, from one another. The elytra,—those horny sheaths which protect the membranous wings,—embrace, in the Coleoptera, the entire upper surface of the long annulated abdomen, and resemble vari-coloured *chlamydes*. But now, look for the elytra of our Forficula. You will hardly believe that they are represented by this kind of abbreviated light-brown jacket, which does not extend below the middle of the back. Do you observe yonder whitish spots? They indicate the tips of the wings, which are longer than their covers. Lift up one of the elytra with your penknife, and you will find that the wing which it partly screens is worth your attention. The fore part (we should call it the *upper*, if the animal walked erect like a man) is straight, and without a fold. Raise it with a pin to see the posterior or lower part. Observe, it curves underneath so as to bind the intermediate portion like a fan. But this flabelliform wing,—of tolerable dimensions when unfolded,—seems intended by the Creative Thought only to mark its unity of plan: the earwig does not fly,—it secures its food by crawling.

The elytra and the wings, inconspicuous as they are, produced so great an impression on the early naturalists, that they made them the principal characteristics of an entire order of insects. De Geer, a celebrated Swedish naturalist, named them the *Dermaptera* (from δέρμα, skin, and πτερόν, wing), in allusion to the transparent skin-like appearance of the elytra. This name, though adopted by Kirby, has not been preserved. A French entomologist suggested the designation which is now in use,—*Orthoptera* (from ὀρθός, straight, and πτερόν),—referring to the manner in which the wings are folded underneath the elytra.

Here we must pause to recapitulate for the benefit of our younger readers, and to avoid confusion, the various orders into which the insect world is divided.

1. *Aptera* (from α, without, and πτερόν, a wing),—wingless. Examples—Flea, louse, chigo.

2. *Diptera* (δίς, two, and πτερόν),—two-winged. Sub-divided into Nemocera, having six-jointed antennæ; Brachycera, having three-jointed antennæ. Examples—Gnat, tipula; May-fly, gad-fly.

3. *Hemiptera* (ἥμι, half, and πτερόν),—half-winged. Sub-divided into Heteroptera, with wings of different textures; Homoptera, with wings of one substance. Examples—Land-bug, water-bug; cicada, lantern-fly.

4. *Lepidoptera* (λεπίς, a scale, and πτερόν),—scaly-winged. Examples—Tiger-moth, butterfly, silkworm.

5. *Orthoptera* (ὀρθός, straight, and πτερόν),—straight-winged. Examples—Earwig, cockroach, locust.

6. *Stymenoptera* (ὑμήν, a membrane, and πτερόν),—membranous-winged. Examples—Bee, wasp, ant.

7. *Neuroptera* (νευρόν, nerve, and πτερόν),—nerve-marked wings. Examples—Dragon-fly, caddis-fly, ant-lin.

8. *Strepsiptera* (στρεψίς, a twisting, and πτερόν),—curled or twisted wings. Examples—Xenos, elenchus.

9. *Coleoptera* (κολεός, a sheath, and πτερόν),—sheathed wings. Examples—Beetle, cockchafer.

The order of *Orthoptera*, with which we are now concerned, is not very well known. The reason is, perhaps, that the insects belonging to it—earwigs, cockroaches, grasshoppers, crickets—are not less disagreeable than useless, so far as man is concerned. Being nearly all of them omnivorous, like man himself, they frequently aid him, very much against his inclination, in the consumption of natural products of every kind.

It has been remarked that the species of great animals are far fewer in number than those of the little. This remark applies with peculiar force to the Orthoptera, which do not include nearly so many small species as the Coleoptera.

The earwig is the type of the tiny group of the Forficulidæ, of which two species only are known to the common world—the *Forficula auricularia* and the *Forficula minor*.

The first species everybody is acquainted with. We have already spoken of its elytra and its wings; but we now say a word upon the two extremities of its body. The two antennæ, which crown the head, are extremely mobile, owing, of course, to the numerous articulations of which they are made up. These are fourteen in number (if we include the base, which is itself composed of two movable parts). In reality, however, there are but twelve; for we ought to eliminate the base,—because, in form and size, it differs greatly from articulations properly so called,—and, at the same time, to regard as one the articulation or joint inserted in it. In fine, I am of opinion, contrary to the general conclusion, that the antennæ of the earwig consist but of twelve articulations, bristling with hairs, and easily counted almost by the unassisted eye. With the help of a microscope, the observer can easily distinguish the large nervure traversing them from top to bottom, and communicating to the antennæ their characteristic sensibility and mobility.

The brownish-coloured abdomen, composed of imbricated rings, forms, in itself alone, upwards of half the body. The animal can move itself in every direction; can bend and twist like a young eel. To the last of its rings, which is larger than the others, are attached the two curved branches of the forceps (*forficula*). These are weapons of defence rather than of attack. At the same time, they are useful as a sexual distinction. The forceps of the male are strongly arched, and furnished with indentations perfectly visible to the naked eye (see Fig. 64, *a*); those of the female are scarcely bent at all, and their indentations can only be seen with the microscope. In numerous individuals, the last ring of the abdomen is provided with four tubercles, one in each side and two in the middle; but this is not a uniform characteristic.

The earwig is a *trimeral* insect; that is, its tarsi are each composed of three joints. Its mandibles are comparatively weak. The moment you touch it, the insect raises, with great quickness, the extremity of its supple body, and endeavours to defend itself with its pincers. The female lays her eggs chiefly in the chinks and crannies of time-worn timber, and the larvæ issuing from them do not differ, in any material respect, from the perfect insect. (See Fig. 64, *b*.)

The small species, known as *Forficula minor*, is not very common. It is about half the size of its better-known congener, and is also distinguished from it by its joints, ten in number,—by its legs, of a very pale yellow,—and by its pincers, which are not only very short, but almost straight, and scarcely marked, even in the male, with any indentations. More, the wings are of the colour of the elytra, and without any white spots. This species is chiefly met with in the spring-time, and then in damp sandy localities, near ponds and rivers.

Another, and still rarer species, to which we may permit ourselves an allusion, has yellow pincers, rather black at the extremity, and garnished inside, towards the middle, with a horny tubercular projection. In the Pyrenees a species has been found which has no wings at all, and has therefore been named *Forficula aptera*.

Our readers will now inquire, What is the use of this curiously constructed animal? Is it not an abomination to the gardener? Well, we admit that it eats up the leaves of his plants, and the petals of his flowers, especially of the dahlia; but, on the other hand, it destroys those far more injurious insects, *thrips, aphis,* and the like.

But it has a peculiar interest for the scientific student from the point of view of what we may call its *muscular dynamometry*,—its power of traction, which is far superior to that of our strongest quadrupeds.

Do you doubt the truth of this assertion? Try, then, the following experiment.

Fasten to the insect's pincer, or forceps, with a thread, a halfpenny, which will weigh about two grains, while the weight of its body, on an average, will not exceed five centigrammes. Give the insect free course over a sheet of paper, and you will see it drag along the coin like a light chariot. Our animal is, therefore, capable of drawing a burden fifty times heavier than its own body. A man of eleven stone would, in the same ratio, be able to drag 7700 lbs. Neither man nor horse can enter, in this respect, into competition with the earwig. If all the members of the animal kingdom were classified according to their power of traction, it is probable that the post of honour at the top of the list would be occupied by our despised *Forficula auricularia*.

The idea of a *muscular dynamometry* of insects is not so new as one might be tempted to think it. From time immemorial men have been struck, without being able to account for it, by the enormous disproportion existing between the weight of a flea and the force or energy displayed by its extraordinary bounds. Hence the popularity of a recent exhibition in London of Performing Fleas. Pliny, eighteen centuries ago, asserted that the muscular strength of the ants exceeded that of all other animals, if we compared the burden they were able to carry with the diminutiveness of their bodies. "*Si quis comparet onera corporibus earum, fateatur nullis portione vires esse majores.*" *

In the seventeenth and eighteenth centuries, this interesting

* Pliny, "Hist. Nat.," xi. 36.

question was taken up by Borelli, Lahire, Buffon, and Gueneau de Montbeliard. Recently it has been revived, with much ability, by Felix Plateau, whose experiments have proved that the insects, in comparison with their weight, possess an uncommon muscular force, far beyond that of vertebrate animals; that in the same group of insects this force varies in different species; and that in the small species it is often of astounding energy.

The muscles are enclosed in solid sheaths (so to speak), which constitute the jointed limbs of insects, and the thickness of the sides of these sheaths seems to decrease in ratio with the size. No relation, therefore, exists between the stature of individuals and the volume and strength of their muscles. A giant may be weaker than a dwarf. Here is another mystery for science to reveal!

But we must take leave of our earwig. Its English name is derived by some authorities from *ear*, and the old English *wiega*, a worm or grub,—identical with the German *oberwurm*, and based, of course, on the fiction which we have already exploded.

Newman, however, suggests a somewhat different name, and, consequently, a different etymology:*—"The shape of the hind wings," he says, "when fully opened, is nearly that of the human ear; and from this circumstance it seems highly probable that the original name of this insect was *earwing*." But we cannot agree with Mr Newman.

* Newman, "Introduction to the History of British Insects."

It remains to be added, that the female earwig sits upon her eggs, and hatches them like a hen; and like a hen, too she gathers her young around her with evident affection.

FIG. 65.—Landscape.

BOOK IV.

AUTUMN.

"Where are the songs of spring? Ay, where are they?
　　Think not of them; *thou* hast thy music too,
　While barred clouds bloom the softly dying day,
　　And touch the stubble-plains with rosy hue;
　Then in a wailful choir the small gnats mourn
　　Among the river shallows, borne aloft
　　　Or sinking as the light wind lives or dies;
　And full-grown lambs loud bleat from hilly bourn;
　　Hedge crickets sing; and now with treble soft
　　The redbreast whistles from a garden croft,
　　　And gathering swallows twitter from the skies."
　　　　　　　　　　　　　　　　—KEATS

"It was a fair and mild autumnal sky,
　And earth's ripe pleasures met the admiring eye,
　As a rich beauty, when her bloom is lost,
　Appears with more magnificence and cost."
　　　　　　　　　　　　　　—CRABBE.

CHAPTER I.

WHAT MAY BE SEEN IN THE HEAVENS.

"The contemplation of the works of creation elevates the mind to the admiration of whatever is great and noble, accomplishing the object of all study, which is to inspire the love of truth, of wisdom, of beauty, especially of goodness, the highest beauty, and of that supreme and eternal Mind which contains all truth and wisdom, all beauty and goodness."—MARY SOMERVILLE.

TO discern the luminous point which should guide us in the shadows of the infinite, is the gift of genius. The first to discern this point in astronomy, the illustrious Kepler thereby succeeded in formulating those laws, or rather rules, by which the movements of the stars are regulated. How did he succeed? How did he arrive at a goal so much to be desired? By intelligence in full possession of itself. It was by abstracting his thoughts from all systematic conceptions, —the shackles of science; it was by defying the traditional authority which had so long enslaved men's minds; it was by interrogating nature, which leaves all liberty to her interrogator, that Kepler was able to de-

T

serve and win the glorious title of "legislator of the heavens," —a title which we must not, however, understand too literally; it bears witness only to the power of intellect.

Let us attempt, at a modest distance, to proceed like Kepler; let us make astronomy without troubling ourselves concerning astronomers. This is the sole means of seizing the luminous point which should guide our steps.

The movement, in virtue of which every star performs the circuit of heaven in four-and-twenty hours, is incessantly reproduced in a uniform and a constant manner. The acquisition of this first fact, simple as it seems, was a somewhat laborious task, and undoubtedly dates back to a distant antiquity. But now comes another fact, where observation demands the closest mental attention, and which is of a more recent discovery.

To comprehend it clearly, let us first call to mind that the moment when the sun crosses the Equator,—whether to return into the northern hemisphere (at the *spring equinox*), or into the southern (at the *autumn equinox*),—is instantaneous. More than one way exists of determining this moment exactly; but here we need not enter upon the subject.

Is the interval of time occupied by the sun in travelling from the spring to the autumn equinox equal to the interval which our luminary requires to pass from the autumn equinox to the vernal?

A singular question, you reply. Who, indeed, would venture to maintain that the number of days, hours, minutes, seconds, was not exactly the same in the one case as in the other?

Well, the period is *not* the same; and, therefore, merely to propound *this* question was a masterpiece of genius. For no ordinary intellectual audacity was needed to doubt the reality of the supposed perfect circle which the sun apparently describes—according to the recognised authorities—in its uniform progress around the globe; that globe believed by all the early astronomers to be imperturbably and everlastingly situated in the centre of their thrice-sacred geometrical figure. This dogma being accepted as infallible, there was every evidence that the two intervals of time, which divided the astronomical year into two moieties, would be of equal duration. It did not occur to the mind of any one of the faithful that the sojourn of the sun, in his circular and uniform movement, might be longer or shorter in the northern than in the southern hemisphere.

What, then, was the name of the audacious innovator who ventured upon putting forth so revolutionary a suggestion?

It was Hipparchus. At least it was he who, confidently relying upon his observations, was the first to affirm that the sun remains longer in the northern than in the southern hemispheres; or, more accurately speaking, that its passage from the spring to the autumn equinox occupies 187 days, while from the autumn to the spring equinox the duration of its course is only 178 days 6 hours (nearly). The year of $365\frac{1}{4}$ days—that is, the Egyptian year, which was universally adopted by the ancient astronomers—was thus discovered to be really divided into two unequal portions, although, *theoretically*, the sun ought to occupy exactly the same space of time in passing from the spring to the autumn, as from the autumn to the spring equinox.

The fact pointed out and attested by Hipparchus had an influence which he never anticipated on the progress of science. In opposition to all the systems previously designed by man, it followed, in the first place, that the movement of the sun, in relation to a mean movement, must sometimes be accelerated, sometimes be retarded ; that the solar arc described in a given time would be greater in winter than in summer.

Astronomers who, trammelled by particular theories, were unable and unwilling to accept of any new light, immediately hastened to raise, as is invariably the case with those who defend a bad cause, a subsidiary and damaging question. They asked whether those inequalities of the sun's movement were real, or only apparent; whether they were more than a mere optical phenomenon, arising from the sun's position *vis-à-vis* to an observer placed on the earth's surface. And they unhesitatingly pronounced in favour of the appearance, and against the reality.

But man, says an old adage, is always punished after the manner of his sin. Dogmatic and obstinate authority involved our anti-revolutionary astronomers in fresh complications. Such is the case, too, very frequently, in the domain of theology!

Does the sun—the sun as each of us beholds him—ever

change his size? Does he ever shrink in his majestic proportions? Is the magnitude of his broad golden disc ever lessened?

Assuredly this new question, which was not less audacious than its predecessor, did not come—there is sacrilege in the thought — from the conservative areopagus of all ancient doctrines; the learned areopagus, or supreme tribunal, which had erected into a dogma the circular orbit and uniform movement of the sun around the earth, the centre of the universe! It could only have been suggested by some unworthy heterodoxical disturber of men's minds,—his name, alas! has not been handed down to us,—who had dared to look upon the heavens, and learn from their bright and beautiful face, without a master, the A B C of science. This "pestilent heretic" had, probably, remarked one of the commonest phenomena connected with the celestial bodies, which astronomers hitherto had not deigned to notice.

Undoubtedly, dear reader, you will have been more than once impressed by the appearance of the solar orb, when obscured in one of those mists so frequent towards the end of autumn :—

> "Cold grew the foggy morn, the day was brief;
> Loose on the cherry hung the crimson leaf;
> The dew dwelt ever on the herb; the woods
> Roared with strong blasts, with mighty showers the floods."

Such a day, in this sad season of the year, when the glory of the woods is rapidly departing, and from the swollen streams and dewy pastures the vapours ascend in a dense whirl of clouds, is of frequent occurrence ; and on such a day, the solar sphere, as it struggles through the screening mists, seems like the face of the moon at its full, when slightly veiled. Your eye

Fig. 66.—" When the glory of the woods is rapidly departing."

rests upon it without pain. And as observation sharpens your mind, you put to yourself the natural question, Is not the sun farther from the earth at this epoch when it affords us the least heat, than at that period of the year when its vivifying power is greatest? I think it is obvious that, to the unexperienced, such a method of explaining the cold of winter and the heat of summer by the variation in the distance of our great solar luminary would naturally occur.

But the demon of certainty—an excellent demon, whatever the orthodox may say—is present, to stimulate us all. You may have just formed your theories, you may cite your traditional authorities, but these will not satisfy our awakened curiosity. We ask for demonstrations, for irrefragable proofs drawn from the Bible of Nature. We will listen to no oracles but those which are confirmed by the voices of God's second revelation.

Therefore, men required to be assured that the sun was really nearer to us in summer than in winter. For this purpose, it was requisite to make, at the beginning of summer, an observation analogous to that which had been made at the beginning of winter, and afterwards to compare the apparent magnitudes of the solar disc at these two opposite periods of the year.

Behold us, then, at work. You are perfectly tranquil as to the result; for you are persuaded beforehand that the sun *must* be farther from us in the cold season than in the hot. You regard this as a self-evident truth, like an axiom of Euclid's.

But Nature is a great magician; she contrives the most dramatic surprises for the mind which takes the trouble to interrogate her in all simplicity and without dogmatic pretensions.

What a *coup-de-théâtre* it was for the observer who first established experimentally that the apparent diameter of the sun is greater in winter than in summer—that we are *nearer* the sun in the cold season, than in the hot!

On more closely examining a result apparently so paradoxical, man discovered that the angle which subtracts the sun, as seen from the earth,—the *visual angle* which gives the sun's apparent diameter,—varies necessarily throughout the year. Thus, the semi-diameter, or radius, which on the 24th of June equals 15′ 45″, will, a month later, have increased one second (15′ 46″); on the 2d of August will equal 15′ 47″; on the 2d September, 15′ 53″, and so on. We put the exact measurements before the reader in a tabulated form :—

LENGTH OF THE SUN'S RADIUS.

On January 21,	16′ 16″	On July 24,		15′ 46″
,, February 25,	16′ 10″	,, August 3,		15′ 47″
,, March 31,	16′ 1″	,, September 2,		15′ 53″
,, April 30,	15′ 53″	,, October 2,		16′ 1″
,, May 30,	15′ 47″	,, November 6,		16′ 10″
,, June 24, minimum,	15′ 45″	,, December 21, maximum,	16′ 17′

We do not trouble the reader with the fractions of a second, which indicate the quantity of the apparent increase of the radius from the end of June to the end of December, and its apparent decrease from the beginning of January to the end of June.

A glance at the above figures shows that the mean of the apparent diameters, all measured at the moment of the sun's passing the meridian, is about half a degree, or 30′; and that—which is sufficiently curious—720 of these mean suns, set one against another, would be required to fill up the contour of a great circle of the celestial sphere. Is it this fact which suggested the idea of dividing the circle into $\frac{720}{2} = 360°$?

Simultaneously with the discovery of the variations of the solar charioteer, it was ascertained that the moments of the sun's passage of the meridian—moments which measure the 365 different positions occupied by the sun in the 365 days of the year—are not separated by equal intervals, or that equal intervals of time do not correspond to the equal angular displacements,—in fine, that the maximum and minimum of the sun's angular velocity coincide with the maximum and minimum of its apparent diameter. Now, remember that the extreme points where the sun experiences its maximum and minimum angular displacement are named, according to Ptolemæus, the former the *perigee*, the latter the *apogee;* or, if we follow Copernicus, the former the *perihelion*, and the latter the *aphelion*.

The aggregate of these facts was known to the ancients; but the manner in which it was sought to explain them merits notice as a specimen of blind attachment to a preconceived system.

Ptolemæus, the organ of the dictatorial astronomy of antiquity, declares, *ex cathedrâ*, that "the inequalities of the sun's movements are only apparent; that they are simply the effects of the position and of the arrangements of the circles in which these movements are accomplished; and that, in this *apparent disorder of the phenomena* (περὶ τὴν ὑπονουμένην τῶν φαινομένων ἀταξίαν), nothing really occurs contrary to their actual immobility (τῷ ὄντι πέφυκε συμβαίνειν οὐδὲν ἀλλότριον αὐτῶν τῆς ἀϊδιότητος)."

Now, according to this dogmatic immutability, the straight lines, or radii, which proceed from the revolving star to the centre of the circle, would describe "equal angles in equal times." This is exactly the contrary of the result obtained, as we have seen, by careful observation.

But this difficulty no more embarrassed the great pontiff of astronomy than a conscientious scruple would perplex the author of a theological dogma. Listen to him :—

"The true cause of these apparent irregularities is explained by two very simple hypotheses. Either the one or the other would account for the phenomena. In fact, if we suppose the movement to occur in a circle described around the centre of the world, and in the plane of the ecliptic, so that the point whence we are looking corresponds with this centre, we must admit either that the planets make their movements equal in non-concentric circles, or that, if these circles are concentric, it is not simply in these circles that they move, but in others, called *epicycles*, carried through the concentric."*

Examine Fig. 67. Here A B G D represent the ecliptic. E its centre, and A E G its diameter; Z H T K is the epicycle, in which the planet moves uniformly around the

* Ptolemæus, "Syntaxis Mathematicalis," iii. 3.

centre A, while the epicycle uniformly traverses the circle A B G D. Now, suppose that the star has arrived at H; it would appear to an observer at E to be more advanced by the uniform movement of all the arc A H; if it be at K, it would appear, on the contrary, to be less advanced by all the arc A K. At Z the star would appear more distant, and at T, nearer than if it were at A.

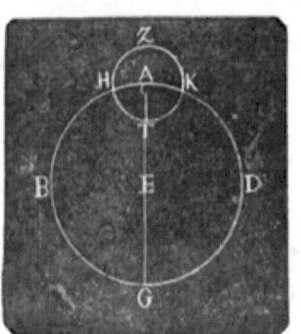

FIG. 67.—The Circle and the Epicycle.

To explain the other phenomena, such as the stations and retrocessions of the planets, recourse was again had to the epicycles or deferred eccentric circles. By multiplying these it was possible to account for all the angular inequalities in the movements of a planet. It is of importance to note this point, in order to show how very dangerous it is to trust absolutely to mathematics in our search after the truth; that science which, by the certainty of its demonstrations, nourishes our intellectual pride, and may, therefore, occasionally lull the mind into a false security. The theory of epicycles, from a mathematical point of view, was irreproachable, and it sufficiently accounted for the facts which threatened to overthrow the dogma of circular orbits and uniform planetary movement.

But by degrees, as observations grew more accurate and

comprehensive, these and other theories, however fine in appearance,—*teres atque rotundus*,—gradually disappeared, if fundamentally erroneous. By the invention of micrometers, we were enabled to measure more exactly than had formerly been possible the variations of diameter or the modifications of distance, and afterwards to compare them with the changes of velocity. From this comparison it results that the latter are not greater than is compatible with the alterations of distance indicated by the variations of diameter; in a word, that the hypothesis of epicycles is decidedly insufficient to account for all the inequalities detected by careful investigation.

Kepler was the first to break the charm which had held captive the mind of astronomers, including even Copernicus and Tycho Brahé. Ptolemæus had considered the mean positions of the stars to be real. Kepler, strong in his researches, declared that they were but a factitious mode of calculation by which the *true* positions might be ascertained; that the mean movement is simply an artifice representing the star's place, *if* no inequality existed; in fine, that we must take the movements as they are in nature,—the true movements, given by observation,—and not the mean movements, deduced from an erroneous hypothesis.

This declaration of principles met, at the time, with the hostility of all astronomers of any reputation, but it has become the starting-point of the discovery of the laws on which the whole edifice of astronomy reposes. Had Kepler, however, been left to depend entirely on his own resources, he might, perhaps, have never completed his task. A fortunate circumstance brought him an unexpected ally. Tycho, having taken refuge in Bohemia, sent for the young astronomer (Kepler then was but twenty-nine years old), to assist him in the composition of the "Rudolphine Tables."*

"This," says Kepler, "was a providential interposition. I repaired to Bohemia early in the year 1600, in the hope of learning the correction of the eccentricities of the planets. Perceiving that Tycho made use of a mixed system (which made Mercury and Venus revolve around the sun, and all these planets, with their companions, around the earth), I asked his permission to follow out my own ideas. It was the will of Providence again, that we should occupy ourselves with Mars. My whole attention, therefore, was directed to this planet: and it is through the movements of Mars we must obtain our insight into the secrets of astronomy, or remain ignorant of them for ever (*ex Martis motibus omnino necesse est nos in cognitionem astronomiæ arcanorum venire aut ea perpetuo nescire*)." †

Why this preference given to Mars? In the first place, because, among all the planets then known, it was Mars which, in its movement round the sun, departed most from the circle; next, its orbit approaches nearest to the earth's; the earth is

---

* Certain astronomical calculations, so called because begun under the patronage of Rudolf II., Emperor of Germany (1576–1612).

† Kepler, "Astronomia Nova, seu de Motibus Stellæ Martis," p. 53 (ed. 1609).

very near to Mars when she passes between that planet and the sun,—that is to say, when she is in *opposition*, while she retires from it triple the distance when in *conjunction*,—that is, when the sun is between her and Mars. Hence arise certain variations of aspect, particularly adapted to make manifest the form of the orbit, and the law of the real movement of the "red planet, Mars." As for the other planets, as far as they were then known, their orbits differ so little from the circle, that the nature of the curve which they describe in reality would never have been exactly recognised by any inexperienced star-gazer.

For these reasons Kepler regarded as providential the choice he had been led to make of Mars at the outset of his astronomical career. Before the close of 1601, Tycho died, bequeathing to his young fellow-worker a treasury of observation. Thenceforth Kepler undertook to finish without assistance the famous Rudolphine Tables. They cost him five-and-twenty years of assiduous labour. Looking upon Tycho's observations, because of their exactness, as "a gift from the Divine Goodness," he employed them, in the first place, as a test of the old hypotheses of planetary orbits and movements. Let us do our best to grasp the range and bearing of this part of his work.

In the system of Copernicus, which Kepler ardently adopted, the earth revolves around the sun. Now, observation having shown that the sun remains seven or eight days longer in the northern than in the southern signs of the Zodiac, we must of necessity admit that the sun, instead of being situated in the centre of the terrestrial orbit, occupies a point *outside* that centre, in such a manner that the earth must sometimes be nearer to, and sometimes farther from, the sun. The distance by which it departs from the centre of its orbit, which Copernicus, like the ancients, supposed to be circular, is called its *eccentricity*.

Astronomers were long preoccupied with the idea of seeking in this *eccentricity* a point where the movements should appear equal. This point was the centre of the *equant*,—a name given to the eccentric circle described from the point of equality or from the centre of the mean movements.

Now, let us recall the principal condition of the problem which Kepler had undertaken to solve. This condition required that the straight line drawn from the centre of our globe to the centre of the sun,—in a word, that the *vector radius*, as it is called, should describe around the sun certain angles, whose variability should agree with the results of observation.

Starting from this point, Kepler found that, for certain positions of Mars (in the *aphelion* and *perihelion*, corresponding to the *minimum* and *maximum* of velocity), the centre of the orbit, always supposing it to be circular, divided into two equal parts (or *bisected*) the total eccentricity: in other words, that it exactly occupied the middle between the centre of the eccentric and the *equant* of Ptolemæus; but it did not appear to him necessary to bisect it in other positions, intermediate between those of the aphelion and the perihelion. He established that the differences in longitude amounted to eight or nine minutes. Now, observations so exact as those of Tycho were altogether incompatible with such great error.* Therefore, the geometrical hypothesis which gave these errors was false; the orbit of Mars could not be a circle, and to save these eight or nine minutes, furnished by observation but in disaccord with theory, it would be needful to recommence all the calculations of astronomy. This conclusion, not less legitimate than daring, supplied Kepler with the first decisive step in the task he had undertaken.

This is not the place to relate all the essays and miscarriages through which this man of genius passed before finally completing his discovery of the rules that bear his name. But we may put before the reader the construction which led to them.

On a sheet of paper let us mark down by a point (Fig. 68) the place occupied by the earth in relation to the sun.† From this point o, we draw a right line terminating at a, the sun's noon-day position (for example, on the 1st of January); the succeeding lines shall touch upon a' a'', which the sun occupies successively after the same interval of time (twenty-four

---

* The reader should turn to Kepler's immortal work, "De Motibus Stellæ Martis," for the manifold attempts of the astronomer to bring calculation into agreement with observation. In every page is revealed what has been finely called "the passionate patience of genius."

† There is certainly no exaggeration in comparing the earth to a point, since the diameter of the sun is 112 times that of the earth, and its mean distance a little more than 12,000 terrestrial diameters: the earth is but a microscopical point in space, if we compare it to the place occupied by the central luminary.

hours, or the exact duration of the earth's rotation on its axis);—and let us continue after this mode until the sun has accomplished, by its own proper movement from west to east, the whole circuit of the heavens, traversing 360 degrees in the space of a year. If we ascribe to the radius oa a certain length, corresponding to a definite solar diameter, the lengths of all the others, corresponding to the variations of the same diameter, will depend upon that of the first, which, for facility of calculation, we suppose to be divided into one thousand parts.

FIG. 68.—Diagram for Kepler's Laws.

After having thus allotted to each straight line its approximate length, let us join their extremities by a curve. What do we see before us? A geometrical figure widely different from a circle, for the diameters (*i.e.*, the straight lines passing through the centre) are far from being equal. The figure is an ellipse.

If now we pass from the appearance to the reality, o will be the sun, and $a\,a'a''$, $m\,m'$ will indicate the terrestrial orbit, or the points of the curve successively occupied by the earth in movement. The moveable straight lines, free at one extremity, and at the other attached to the centre of the sun, are called the *Vector heliocentric radii*. By the help of this construction, you see that the point occupied by the sun is beyond or without the centre; this eccentric point is the *focus* of the ellipse, and the distance from this focus to the centre, its *eccentricity*. The extremity of the major axis, the nearest to the focus, is the *perihelion*, and its farthest extremity the *aphelion*. The difference of the angles formed by the vector radii indicate the inequality of the movements: to the greatest angle, the perihelion, corresponds the *maximum* of velocity ($a\,a'a''$), just as to the smallest, or aphelion, corresponds the *minimum* ($m\,m'$); the other angles mark the velocities intermediary between these two extremes. We have thus before us a series of triangles with their apices at the focus of the ellipse, and their bases on the contour of the curve.

But these latter are not sufficient for the mind, whose principal function lies in seeking unity among the variety of phenomena.

In what way are the variations of distance connected with the variations of velocity? What is the simplest expression of their relationship? These are questions which naturally presented themselves to Kepler's inquiring intellect. By dint of immeasurable patience, and recommencing more than once the same toil, this great astronomer discovered that the variable arc traversed by the earth (or, in appearance, the sun), in four-and-twenty hours, multiplied by one half the corresponding vector radius, is a constant quantity: is the product which, as elementary geometry teaches, gives the surface of a triangle. And, in fact, look at the matter carefully: the *vector radii* form triangles whose base is the arc traversed in the same interval of time, and whose apices rest upon the centre of the sun (or, in appearance, the observer, or the centre of the earth).

To fix these ideas thoroughly in our minds,—and a superficial knowledge is worse than useless,—let us imagine to ourselves a man holding horizontally extended a tube of a certain length, capable, like a telescope, of being lengthened or shortened at pleasure; and let us fancy him pivoting upon himself, in such a manner that he sweeps, every minute, exactly the same area or same quantity of surface, while varying perpetually the swiftness of movement and the length of the tube; this "ideal man" will have solved the problem whose solution is inscribed, in ineffaceable letters, on the machinery of our globe; he will describe around him an ellipse, of which he himself occupies one of the foci.

By this method of investigation and deduction, Kepler succeeded in breaking up the traditionary authority of the circle and of uniform movement. He broke it up for ever by two of his celebrated laws, which may be rendered in the following terms:—

1st, *The orbit of the earth, as well as the curves described by the other planets, are ellipses, one of whose foci is represented by the sun;*

2d, *The heliocentric vector radius of a planet describes around the sun areas equal with the times; or, in other words, the surfaces described by the vector radii, in equal times, are also equal.*

The ancients had looked for equality in the movements of

planets traversing the circumferences of circles: they were mistaken. It is true this equality exists; only, not where they supposed. If they had sought it in the surfaces described by the vector radii, they would have anticipated Kepler's discovery of the laws which govern our world.

But their astronomical dogmas prevented them from seeing the path which led to this great discovery.

Hence we may conclude that Dogma is an evil thing.

CHAPTER II.

WHAT MAY BE SEEN UPON THE EARTH.

"I care not, Fortune, what you me deny;
You cannot rob me of free Nature's grace;
You cannot shut the windows of the sky
Through which Aurora shows her bright'ning face;
You cannot bar my constant feet to trace
The woods and lawns by living streams at eve."

OF all the strata composing our planetary mass, the most important, so far as man is concerned, is, at the same time, the most superficial; for it is here that all the phenomena of life transpire. Our vegetable earth is the great laboratory in which are prepared all the solid, liquid, and gaseous aliments necessary for the nourishment of animal life. It is on the surface of the globe that men play their various parts. And why? Can it be for no other purpose than to modify, in some degree, its aspect, that they occupy the terrestrial surface? One would be tempted to think so on consulting what these majestic *bimanes* pompously designate their "Universal History." Regions formerly blooming with fertility,—gay with gardens, and orchards, and meadows,—musical with brooks, and glorious with harvest,—are now uncultivated and barren. Monuments which seemed adapted to defy the winds and the rains, and the corroding touch of the years, lie shattered in ruins; and with them the once populous cities and the once mighty empires of which they were the pride. The jackal howls among the broken columns of Tadmor; the sand-drifts have accumulated above the splendour of Memphis and Thebes. With their stones other monuments are raised, other cities are embellished, and other empires, which, in their turn, undergo the same unalterable fate: a perpetual relation of human forms, in every respect comparable with that which transpires in the bosom of the prolific earth, our common mother and nurse.

But why do men wander so far from the straight way? Why do they their best to ensure each other's unhappiness? They seem, alas! ignorant of the tendency of their actions, while attaching themselves to things transitory, and despising things imperishable. These, indeed, they would utterly ignore; they would live, like the brutes, unconscious of their destiny, if, at the bottom of their indestructible conscience, there did not prevail a glimmer of light, though more or less eclipsed, if they did not all feel themselves attracted, if they did not all irresistibly gravitate, some more quickly, others more slowly, towards the sun of eternal truth and justice. Instead of moving with sidelong sinuous pace, instead of taking ninety-nine steps backward for every one hundred taken in advance, they would all march onward in the way of progress; were it not that they pass their time in clipping their own wings; were it not that, to bend their heads the better—*Veluti pecora ventri obedientia*—they check the aspiring flight of that thought which would soar beyond the present; in a word, were it not that they lay a sacrilegious hand—unfortunate wretches!—on that which God Himself has respected in His creature—Liberty! The doubt which perplexes us as to the great problem of our destiny,—the doubt which allows so much latitude to the workings of our conscience,—does it not indicate the path we ought to follow? Should not men regard their freedom with peculiar reverence, when the Divinity they invoke has mercifully refrained from fettering it? Creatures of a day, who live as if you would never die! the contradictions and the miseries of which you so incessantly complain, are your own work. Help, help yourselves, by the development of your faculties, by the cultivation of your heart and mind, for herein you shall see the law and the prophets. Barren lip-service is nothing better than blasphemy!

But let us return to the ground which we tread, and where our life-companions are the animals and the plants.

The uppermost stratum of our globe undergoes the direct action of the light and heat of the all-vivifying "orb of day." This action, very unequal in its effects, and most important to understand, has scarcely been touched as yet by scientific research. Our geologists, having been more busily engaged with the inside than the outside of the earth, have broached certain plausible theories—for the most part of a very dubious character—respecting the central fire, Plutonism and Neptunism, the stratification of the planets, the formation of mountains, valleys, and basins. Our mineralogists, thinking far less of the chemical molecular constitution of the different formations than of their crystalline constitution, have minutely studied the physical qualities and geometrical forms of the integral parts of the rocks; but neither have condescended to direct their inquiries to the layer of soil trodden underneath their feet. Yet this very layer of arable earth, to which all bodies must return after death what they have taken from it during life,—this much despised *humus*, furnishes all our agricultural products, the very foundation and support of our material existence.

To touch industrial occupations—to meddle with trade, commerce, or agriculture—is unworthy of Science! Such is the silly cry of the many distinguished savants who pride themselves on what they call their "freedom from selfish considerations."

Be it so; but then you ought surely to be consistent, and never regard science as a profession or a bread-winner.

On the Chemical Action of Light.

It is no easy study to investigate the modifications and chemical effects which the terrestrial surface is capable of receiving or undergoing, either from the direct rays of the sun, or from diffused light. It requires new methods of inquiry, —methods frequently of extreme delicacy, as the labours of

Bunsen and Roscoe, of Kirchhofer and Tyndall, have abundantly demonstrated. Let us here confine ourselves to establishing the fact that the chemical action of light varies according to the geological constitution of the soil,—according to the diurnal and annual obliquity of the solar rays,—according to the hours of the day,—according to the latitudes and seasons. The maximum of effects is manifested about the times of the solstices.

For the better co-ordination of these phenomena, might we not, as has been done in regard to the distribution of heat over the terrestrial surface,* link together by lines the points of equality? We should thus create an aggregate of *iso-photo-chemical* lines,—diurnal, mensual, and annual,—of incontestible utility for the progress of general physics and meteorology, which are still in their infancy.

But to realise this magnificent programme, the union and agreement is necessary of scientific men in every region of the globe; an ideal, therefore, as yet, is very far from being attained.

The Action of Heat.

The earth is subject to the influence of two different sources of heat. One, like the arterial blood, strikes from the centre to the circumference: this internal heat it is which has been stored up since the unknown epoch when our globe was nothing more than an incandescent nucleus, surrounded by condensable vapours. The other, like the

---

* Reference is here invited to Humboldt's *thermal* and *isothermal* lines. See Dr A. K. Johnston's "Physical Atlas," and Humboldt's "Kosmos."

venous blood, flows from the circumference towards the centre: this is the solar heat which the earth continues to receive through its crust.

The first of these sources lies beyond the domain of experiment. It has been the object of numerous hypotheses and diverse speculations, with which we shall not here concern ourselves. The second source is alone accessible to our investigations, and yet the network of *isothermal lines* is scarcely defined.

Since the year 1817, when Alexander von Humboldt conceived the felicitous idea of representing by lines the same mean temperatures enjoyed, in a given space of time, by the different regions of the globe, researches of this nature have very considerably multiplied. But these researches—do not forget—refer rather to the temperature of the atmosphere than to the heating of the inferior stratum of that gaseous ocean whose bed or foundation is the terrestrial crust. And it is the penetration of the latter by the sun's calorific rays which we would especially desire to understand. Here, then, a sufficient margin is left for our curiosity.

"If the king, my father, does not rest from his conquests," cried Alexander of Macedon, while still a child, "he will leave me nothing to do when I shall have reached manhood." To such a complaint, be you sure, dear reader, that the conquests made by science will never give rise. Every step in advance is but a step into the infinite: what we have done only shows us the boundless extent of what we have to do.

It is this reflection which must always teach humility to the scientific student, even while he rejoices in the achievements of human patience and genius. He will not despair for he knows that great victories have been won: he will not grow arrogant, for he knows that he is still on the threshold of eternal truth. As Sir J. Herschel has justly said:—"He who has seen obscurities, which appeared impenetrable, in physical and mathematical science, suddenly dispelled, and the most barren and unpromising fields of inquiry converted, as if by inspiration, into rich and inexhaustible springs of knowledge and power, on a simple change of our point of view, or by merely bringing them to bear on some principle which it never occurred before to try, will surely be the very last to acquiesce in any dispiriting prospects of either the present or the future destinies of mankind; while, on the other hand, the boundless views of intellectual and moral, as well as material relations, which open to him on all hands in the course of these pursuits,— the knowledge of the trivial place he occupies in the scale of creation, and the sense continually pressed upon him of his own weakness and incapacity to suspend or modify the slightest movement of the vast machinery he sees in action around him, must effectually convince him that humility of pretension, no less than confidence of hope, is what best becomes his character."*

The temperature of the terrestrial surface perpetually varies

---

* Sir J. Herschel, "Preliminary Discourse on the Study of Natural Philosophy."

under the influence of local as well as general causes. Had this fact been known to the philosophers of antiquity, they would have taken advantage of it to liken the earth to an animal whose skin is more or less sensible of heat, not only according to the difference of the seasons, but according to the different hours of the day.

The diurnal thermometrical variations are those which penetrate the least profoundly into the interior of the soil. At a depth of about five feet they cease to be perceptible. The maxima and minima of the year, however, can be detected at a sufficiently considerable depth. The limit descends as low as 80 to 100 feet. Below 100 feet, the terrestrial stratum is found invariable,—that is, inaccessible to the thermometrical changes of the atmosphere. It is remarkable that the temperature of this stratum differs but little from the mean annual temperature of the air, which, in the latitude of London, is 49°.

The maxima and minima of the yearly heat are propagated very slowly in the earth, and their difference gradually becomes less and less. Thermometers buried 26 feet deep in the ground, mark, in our latitudes, the maximum of temperature only on the 10th of December, and the minimum on the 15th of June. But there are certain elements which we must take into account. Thus, the depth of the invariable thermometrical curve depends both on the latitude of the place, on the conductibility of the strata, and the difference between the highest and lowest temperature of the year.

The less this difference, the more nearly does the invariable stratum approach the surface. Here we have the reason why, in the intertropical torrid zone, where the temperature scarcely varies above two or three degrees in the whole year, the invariable curve is not found more than forty centimetres beneath the surface.

In the temperate zone, lying between the torrid and the frigid zones, the same phenomena assume, apparently, a more complex character; the isogeothermal lines inflect as diversely as the isothermal, and the former are far from running parallel with the latter. And this is easily understood, even without any experiment, for there is no relation between the ever-varied composition of the terrestrial strata, and the much more uniform composition of the atmosphere.

In the frigid zone, the soil remains constantly frozen for an insignificant depth, whatever may be the temperature of the encircling atmosphere. In some regions a stratum of ice and snow eternally reposes on the surface of the soil.

Unfortunately the observations, which require to be undertaken on an uniform plan and under well-weighed conditions, at various points over the whole globe, are as yet far too few to assist us in defining any general currents of heat or cold, whether variable or constant, as prevailing in the lower strata of the gaseous ocean of our planet.

Arable Land.

A more useful picture than that of the isogeothermal lines would be one of all the arable land covering the continents of the Old World and the New,—indicating the composition of this nutritive earth, the nature of the soil on which it reposes, as well as the various kinds of cultivation appropriate in different climates. Here is a work to be achieved,—a work which would benefit the whole human race,—a work differing vastly from the conquests and achievements of too many of those "heroes" the world delights to honour.

> "Peace hath her victories
> More renowned than war."

In this immense task, of which, as yet, not even the outlines have been sketched, particular attention would require to be paid to the *subsoil;* for upon this the success of all cultivation literally depends.

Arable land is the most superficial stratum of the cultivable terrestrial crust; it is this which the plough turns up and subdues; it is this which, properly manured, and enriched by the decomposition of organic matter, furnishes to vegetables their principal nourishment. As it varies in thickness, it necessarily presents

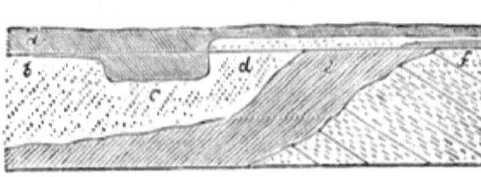

FIG. 69.—*a*, Humus, or stratum of arable earth,—the horizontal line shows the depth reached by the labourer; *b*, subsoil; *c*, subsoil; *d*, arable earth; *e*, humus in an inclined stratum; *f*, humus mixed with subsoil.

one or other of the following circumstances:—1st, The labourer, penetrating the entire stratum of arable earth (Fig. 69, *a*), will strike down to the subsoil (Fig. 69, *b*); or, 2d, he will not traverse the entire stratum (Fig. 69, *c*); or, 3d, after having traversed the entire depth of the humus, he will reach a portion of the subsoil (Fig. 69, *d*); or, 4th, after having gone through both humus and subsoil, he will discover another layer of arable earth, which may be either pure humus, in a thick inclined stratum (Fig. 69, *e*), or humus mixed with the débris of the subsoil.

As for the subsoil, it may, by its composition, completely modify, stimulate, or delay the action of the vegetable mould, however rich this may be in assimilating principles. Thus, where the subsoil is argillaceous, the pluvial waters are arrested by it as by a bed of impervious cement, and render the ground too damp and cold to yield abundant harvests. In such a case subsoil-drainage is the best remedy. But if the earth be porous, the moisture gradually percolates through its various layers, fertilising and warming, communicating to the plants the needful humidity, and assisting in the production of that most glorious of all the scenes of cultivated nature—a corn-field thickly ripe with golden grain. In the poet's "Palace of Art" no finer picture can be seen than this:

> "The reapers at their sultry toil,
> In front they bind the sheaves. Behind
> Are realms of upland, prodigal in oil,
> And hoary to the wind."

Oh! a sight to thank God for, and rejoice in, is the field all aglow with the splendour of the harvest!

Without having recourse to chemical analysis, which is within the reach of a very limited number of persons, *clayey* soils

FIG. 70.—" Behind are realms of upland."

may be distinguished by the vegetable species that most commonly flourish in them: as—

I. *Plants belonging to clayey soils.*—The Queen of the Meadows, *Spiræa ulmaria* (order *Rosaceæ*). Wild Angelica, *Angelica sylvestris.* Common Sorrel, *Rumex acetosa* (order *Polygonaceæ*), and various kinds of Ranunculaceæ, as *Ranunculus lingua, Ranunculus flamma,* and *Ranunculus sceleratus.*

II. *Plants belonging to sandy soils.*—Horny Lotus, *Lotus corniculatus* (order *Rhamnaceæ*). Little Harebell, *Campanula rotundifolia* (order *Campanulaceæ*). Eyebright, *Euphrasia officinalis* (order *Scrophulariaceæ*). *Anthoxanthum odoratum.*

III. *Plants belonging to argilo-calcareous soils.*—Coltsfoot, *Tussilago farfara* (order *Compositæ*). Wild Mustard, *Sinapis arvensis* (order *Cruciferæ*). Buckwheat, *Polygonum aviculare* (order *Polygonaceæ*).

IV. *Plants belonging to a sandy and calcareous soil.*—Broom, *Genista scoparia* (order *Leguminosæ*). Centaury, *Centaurea nigra* (order *Gentianaceæ*). *Galium verum* (order *Rubiaceæ*). The Jacobea, *Seneecio Jacobæa.*

V. *Plants belonging to alluvial and marshy soils.*—Reed, *Arundo phragmites*, *Poa aquatica*, *Poa fluitans*. Rush, *Juncus conglomeratus* (order *Juncaceæ*).

After these different soils have been brought under cultivation, the characteristic species, which we have just enumerated, disappear, and are replaced by other plants, which grow, to all appearance, spontaneously, under the name of *weeds;* but, in reality, spring from germs or seeds too frequently mixed up with the different manures, or spread abroad by the agency of birds or the wind.

In reference to this latter consideration, the diffusion of plants, we shall transcribe an interesting passage from Balfour's excellent "Manual of Botany."

"Some plants," he remarks, "are disseminated generally over

the globe, while others are confined within narrow limits. Some of the common weeds in Britain, such as chickweed, shepherd's purse, and groundsel, are found at the southern extremity of South America. *Laura minor* and *trisulca*, *Convolvulus sepium*, *Phragmites communis*, *Cedium Mariscus*, *Scirpus lacustris*, *Juncus effusus*, and *Solanum nigrum*, are said to be common to Great Britain and New Holland. *Nasturtium officinale*, and *Samolus Valerandi* are very extensively diffused, and they may be reckoned true cosmopolites. They are both natives of Europe, and they occur, the former near Rio Janeiro, the latter at St Vincent. *The lower the degree of development, the greater seems to be the range.* Some cryptogamic plants, as *Lecanora subfusca*, are found all over the globe.

" Man has been instrumental in widely distributing culinary vegetables, such as the potato, and the cereal grains, as well as many other plants useful for food and manufacture. Corn plants, such as barley, oats, rye, wheat, spelt, rice, maize, and millet, are so generally cultivated over the globe, that almost all trace is lost of their native country. They can arrive at perfection in a great variety of circumstances, and they have thus probably a wider geographical range than any other kind of plants.

" As regards these plants, the globe may be divided into *five* grand regions—the region of *rice*, which may be said to support the greatest number of the human race ; the region of *maize;* of *wheat ;* of *rye;* and lastly, of *barley* and *oats*. The first three are the most extensive, and maize has the greatest range of temperature.

"The grains extending farthest north in Europe are barley and oats. Rye is the next, and is the prevailing grain in Sweden and Norway, and all the lands bordering on the Baltic, the North of Germany, and part of Siberia. Wheat follows

FIG. 71.—The Rye District.

rye; it is cultivated in the middle and South of France, England, part of Scotland, part of Germany, Hungary, Crimea, and the Caucasus. We next come to a district where wheat still abounds, but no longer exclusively furnishes bread,—rice and maize becoming frequent. To this zone belong Portugal, Spain, part of France, Italy, and Greece, Persia, Northern

India, Arabia, Egypt, the Canary Islands, &c. Wheat can be reared wherever the mean temperature of the whole year is not under 37° or 39° F., and the mean summer heat, for a period of at least three or four months, is above 55°. It succeeds best on the limits of the sub-tropical region. In the Scandinavian peninsula, the cultivation of barley extends to 70° N. latitude, rye to 67°, and oats to 65°. The cultivation of rice prevails in Eastern and Southern Asia, and it is a common article of subsistence in various countries bordering on the Mediterranean. Maize succeeds best in the hottest and dampest parts of tropical climates. It may be reared as far as 40° N. and S. latitude on the American continent, on the western side; while in Europe it can grow even to 50° or 52° of latitude. It is now cultivated in all regions in the tropical and temperate zones which are colonised by Europeans. Millet of different kinds is met with in the hottest parts of Africa, in the South of Europe, in Asia Minor, and in the East Indies." *

Professor Houston furnishes the following table in illustration of the distribution of *wheat* and *barley*. It also shows the mean temperature which they require :—

BARLEY.

	Winter Mean.	Summer Mean.	Annual Mean.
N. Lat. 62¼° Faroe,	39°	51°	45°
70 Lapland,	22	46	33
67·30′ Russia,	9	46	32
57·30′ Siberia,	0	60	32

* Professor Balfour's "Manual of Botany," pp. 566-568.

		WHEAT. Winter Mean.	Summer Mean.	Annual Mean.
Lat. 58°	Scotland,	36°	57°	64°
	Norway,	23	59	39
	Sweden,	23	59	39
	Russia,	15	60	37
30°	Cairo,	57	88	72
	Macao,	64	82	73
	Rio Janeiro,	68	78	74
	Havannah,	71	82	77
	Bourbon,	71	80	77

"Winds, water, and animals, are also instrumental in disseminating plants. Many seeds, with winged and feathery appendages, are easily wafted about; others are carried by rivers and streams, and some can be transported by the ocean currents to a great distance, with their generating powers unimpaired."

MUSHROOMS OR AGARICS (*Order* FUNGI).

Every year—principally in autumn—we are startled by hearing or reading of cases of poisoning by mushrooms. Erudite connoisseurs, however, who have profited by Dr Badham's book on "Esculent Fungi," do not suffer themselves to be intimidated by these sad narratives, though, unfortunately, they are frequently too well founded; because they know how to distinguish the good from the ill, the true from the false, the edible from the poisonous mushroom. But this security ought not to embolden the inexperienced amateur in risking his life for the sake of a delicacy. It is true, nevertheless, that we frequently see men's lives exposed for something less.

"Look, what a splendid mushroom I have discovered!" a

lady said to me, the other day; a lady who knew something of flowers, but nothing of *cryptogams*. "Take care!" I replied, taking from her hand the supposed prize; "this is the false mushroom; and would suffice, if served up at your table, to poison yourself and all your guests."

I ought here to observe that my friend had narrowly escaped death two years before, through regaling herself with a dish of mushrooms of very dubious character. Among the symptoms which she experienced, and which she described with medical exactness, she particularly dwelt upon the cold sweats, accompanied by a sentiment of undefinable terror, which is nothing else than the dread of death : it was the special symptom of poisoning with an unwholesome fungus.

Let us endeavour to make ourselves better acquainted with this formidable enemy of epicures. You will have no difficulty in finding it in any warm, close season, but especially in spring and autumn. The toadstool thrives indifferently in the shade of all the forest trees, but seems to prefer the oak and birch to the pine and fir. A patch of soft greensward, at the foot of an old oak, and in the neighbourhood of a "brawling stream" or "tranquil pool," will generally be covered with fungi of this description. The numerous synonyms attaching to it show how greatly it has exercised the *classifying spirit* of our naturalists. Some call it *Agaricus muscarius*, as if we should say "kill-fly mushroom;" others, changing only the specific designation, designate it *Agaricus pseudo-aurantiacus*,—which signifies, literally, "false orange," in allusion to the beautiful yellow colour of the *true* aurantiacus. What is certain is, that our mushroom, which can kill men as well as flies, belongs to the genus *Agaric*, so numerous in species that it can be formed into a family, that of the *Agariceæ*. The Agarics, of which the

FIG. 72.—"At the foot of an old oak."

esculent mushroom (*Agaricus edulis*) represents the type, are easily recognised by their more or less fleshy *pileus*, or cap, garnished underneath with *lamellæ*, or gills, which radiate from the centre to the circumference, when the pedicel is in the centre.

But some cryptogamists are unwilling to recognise the *orange mushrooms* (*les oronges*), whether true or false, as Agarics. They place them in a separate genus, the genus *Amanita*, though without informing us where they found the name. Meanwhile, they justify the formation of the new genus by the presence of the white swelling, the *volva*, or wrapper, of the *mycelium*, or spawn, which entirely covers both the true and the false mushroom on its emergence from the earth. Each, then, is an *Amanita*. But now remark their specific difference. The true mushroom, as it develops, ruptures its ovoid wrapper, or volva, leaving the remains entirely at the base of the pedicel; while, in the false mushroom, the débris of the volva are formed, not only at the base of the pedicel, as in the real Agaric, but even upon the red surface of the pileus itself: these are the white irregular warts characteristic of the Amanita, but wholly wanting in the Agaricus. Thus, there are two Amanitas: the *Amanita muscaria*, or "fly agaric," and the *Amanita aurantiaca*, or, as the English botanists call it, *Agaricus Cæsareus*, the imperial mushroom.

This fanciful "study" in nomenclature has the advantage of initiating us into the most essential distinctive characters of the two species in question. However, a few additional details are necessary to complete our history.

The Fly Agaric, or Amanita muscaria.

The species seems to have been expressly created to teach our gourmands the necessity of vigilance; that before enjoying a dainty they must first learn to distinguish, under penalty of death, the poisonous fungus from that which safely and pleasantly tickles the palate. The warning is useful, moreover, as showing that even sensualists are not wholly exempt from the law of work.

The *Amanita*, like most falsehoods, is pleasant to the sight. Its round pileus, of a beautiful orange-red colour, spotted with white warts, and lined with white gills, seems to invite your attention. (Fig. 73.) Strike it down with your stick; the lamellæ or gills underneath resemble the white leaves of a book. Its graceful stalk, ornamented on the upper part by a well-designed necklet, is bulbous in its lower portion; its flesh is dazzlingly white; in short, its entire appearance is attractive. Yet it is a traitor! You little know the depth of its wickedness. Mix a few shreds of its white flesh with a little milk; every fly which drinks of the mixture will, in a few seconds, fall dead, their swollen abdomen bearing testimony to the effect of the poison. And in many districts of Germany, particularly in Thuringia, the peasants make use of it to rid themselves of the swarms of flies which, towards the close of summer, infest their habitations. It is thus that man may frequently turn to his advantage those objects of nature which, at the first glance, appear injurious rather than useful.

FIG. 73.—The *Amanita Muscaria*.

Various experiments have been essayed to test the intoxicating influence of our *Amanita*. Bulliard, author of the "Histoire des Champignons de France," made two cats partake of it. Six hours afterwards, these animals, who are so tenacious of life, were dead.

Haller, in his "Histoire des Plantes Vénéneuses de la Suisse," says that "the *Agaricus muscarius* (*Amanita muscaria*) cannot be eaten with impunity, six Lithuanians having died of it; and in Kamtschatka it has been known to excite deadly attacks of delirium, accompanied by so deep a despondency, that those who have eaten of it would fain fling themselves into the fire, or fall upon their knives or daggers." This, however, we take to be an exaggeration. The truth is, that in Kamtschatka it is used to produce intoxication; and such is its strength, it imparts an intoxicating property to the urine of those who swallow it. When the fungus itself is not at hand, the would-be drunkard frequently resorts to this nauseous potion.

The flesh of the *Amanita* is not yellow. Yet Vicat speaks of poisonings produced by the *Yellow Amanita*. "I had much difficulty," says this physician, "in saving two families of Lausanne, poisoned through eating a very small quantity of mushrooms, which the father in the one case, and the mother in the other, had mistaken for *Agarici Cæsarei*, though they were both esteemed great connoisseurs, and especially in this species; nor had they been once deceived for upwards of thirty years, until they indulged themselves in this delicious but deceitful dish."

These poisonings could have been occasioned only by the *Amanita*. Some varieties exist, in which the under surface of the pileus is yellow, but the flesh is white. To one of these varieties Vicat's anecdote probably refers.

Thus, the *Amanita formosa* of Persoon has pedicel, pileus, and warts of the pileus, of a citron yellow. It is but a variety of the *Amanita muscaria*.

The *Amanita umbrina* of the same botanist (the *Agaricus pantherinus* of De Candolle) has an olive-coloured pileus; its surface, like that of *Amanita muscaria*, is covered with white scales.

The *Amanita solitarius* is distinguished by the size of its umbilical cap,—sometimes depressed in the centre,—which is furnished with a great number of white or pale brown scales; when fully developed, it measures from thirty to forty centimetres in diameter. The pedicel is bulbous, with a membranous ring, white as snow, clasped around it; at the base it is clothed in pellicles, the remains of its scaly volva. The flesh is firm, thick, and white. You rarely meet with more than two or three individuals in the same locality; hence its name of *solitarius*. Bulliard speaks of the flesh as good to eat, when cooked on a gridiron, and seasoned with fresh butter, salt, and pepper. It is possible. But as it is so easily confused with the poisonous species, the author of the "Histoire des Champignons" would have done better to prohibit its consumption, whether eatable or not.

We may now turn to the method adopted by Dr Vicat to save the lives of the two families at Lausanne, who, as we have seen, were poisoned by partaking of the *Amanita*:—

I dissolved, he says, six grains of tartar emetic in a litre of water, and from time to time administered a spoonful to my patients; moreover, I made them swallow floods of warm water, sweetened with a little honey,—that is, a large teaspoonful of honey to a cupful of water.

I had much difficulty to get one of the sufferers, who was sixty years old, to swallow the first few spoonfuls. He was plunged into a lethargic insensibility differing in no respect from complete apoplexy; his teeth were closely set. Those whom I had ordered to administer the mixture had given it up, after several useless attempts; and in all probability the old man would have sunk, had I not had the patience to hold, for some hours, against his teeth the back of the blade of a small silver knife, so as to profit by the few moments when the teeth were a little less firmly clenched. I used some force to make the blade act as a wedge, and after a while opened up a passage to the handle, which, serving as a lever, forced the jaws sufficiently apart to admit the introduction of a spoonful of the emetic. It was not, however, until fully two hours had passed that the patient, having undoubtedly swallowed the necessary dose, began to vomit, with strenuous efforts and frightful cries. This was at midnight. Four in the morning arrived before, after numerous alternations of vomiting and profound lethargy, he began to speak, and then like a man in delirium. After the first vomit, which was inconsiderable, the convulsions of his whole body were so very violent as to require four men to hold him, while I continued to make use of my knife as at first. Nor did I desist until I was satisfied that his stomach had been sufficiently cleansed. After this, I applied two strong blisters to the back of his legs.

As these acted, the purging subsided, and at the end of twenty-four hours it had passed away entirely, the invalid finding himself as well as could be expected after sustaining so severe a shock. The other patients, who were not in so much danger, experienced twitchings and tremblings in the face, which quite disfigured them; the brain seemed a blank; though awake, they felt as in a dream, and their visions were most frightful.

It is evident, from these particulars, that mushroom-poisoning specially affects the encephalic nervous system, and that the best remedies are emetics and antispasmodics. In our present ignorance of what are the poisonous principles in the *Amanita*, we can adopt no other method than a chemical neutralisation.

AGARICUS CÆSAREUS, OR IMPERIAL MUSHROOM.

In this mushroom, for which, as we have seen, the *Amanita* is too frequently mistaken, the inside as well as the outside is yellow; the upper surface of the pileus, which is equally free from scales and warts, is, however, of a reddish yellow, like that of an orange (whence the popular French name, *la vrai oronge*); all the other parts are of a beautiful citron hue. This agaric exhales an agreeable odour, combined apparently of the scent of the vanilla and the truffle. It decomposes rapidly, and when in a state of advanced putridity, the fragrance I speak of is succeeded by—well, by a fearful stench! When young, and still completely covered with its wrapper or *volva* (this, in the *Amanita*, is imperfect), it is very like a hen's egg which has been partly buried in the ground so as to expose only the larger end. It seems partial to solitude; more than four or five are seldom found in the same locality. Moreover, in autumn it affects the same habitats as the *Amanita*,—which is unfortunate.

FIG. 74.—The Imperial Mushroom.

It would seem that our imperial mushroom was specially appreciated by the ancients, and it is said that Nero pronounced it a dish fit for the gods. In this circumstance originated the scientific name which has now become popular, and which was first applied to it by the cryptogamist Fries, *Agaricus Cæsareus.*

The *Boletus* of Pliny appears to have been our Agaricus, and not one of our *Boleti*, which are easily recognised by the numerous tubercular projections covering the under part of the pileus. In proof of this I would point out that the Roman naturalist, after speaking of the *Boleti* as genuine delicacies, immediately inveighs against them as dangerously poisonous. He relates that it was with one of these, or rather with one of the false mushrooms so easily mistaken for the true, that Agrippina poisoned the Emperor Claudius, to secure the imperial crown for her son Nero.*

The virtues of the mushrooms have been sung by Juvenal and Martial. The latter accurately distinguishes the true from the false, when reproaching Cæcilianus with his gluttony.

* Pliny, "Hist. Nat.," xxii. 46.

"Ah, you are used to devour your *Boleti* alone, in the face of your invited guests; eat then, the *Boletus* which Claudius ate!"

> "Dic mihi, quis furor est? Turba spectante vocata,
> Solus boletos, Cæciliane, voras.
> Quid dignum tanto tibi ventri, gulaque precabor?
> Boletum, qualem Claudius edit, edas."*

The best known and most valuable species of Agarici may be briefly enumerated :—

Agaricus campestris, or *Common Mushroom*, found in nearly all temperate regions: pileus convex, and white, with a tinge of brown; thick set on the under side with dark brown gills; stem firm and fleshy, and surrounded by a white membranous ring.

Agaricus Cæsareus, or *Imperial Mushroom*, the *Kaiserling* of the Germans, already described.

Agaricus deliciosus, or *Orange-milked Agaric*, found in coverts of fir and juniper; pileus viscid, orange, and upwards of four inches broad; gills and juice of a fine orange colour.

Agaricus procerus, or *Parasol Mushroom*, found in the shade of trees, on meadows with a sandy soil; stem from eight to twelve inches high, with a thick spongy ring; pileus bell-shaped, and covered with brown scales.

Agaricus Virgineus, or *White Field Agaric*, found in rich moist

---
* Martial, "Epigram," i. 21.

pastures; pileus whitish, and convex; gills of a light chocolate shade; stem nearly two inches broad.

Agaricus eburneus, or *Ivory Mushroom*, found in beech woods; grayish-yellow pileus; broad gills; stem long and scaly.

Agaricus Georgii, *St George's Agaric*, or *Whitecaps*, found in moist pastures, and in the shelter of old barns, farmhouses, and churches; flesh yellow; gills yellowish-white; pileus twelve to eighteen inches broad; the least valuable of British species of agaric, but useful in ketchup-making.

Agaricus oreades, *Fairy-ring Mushroom*, or *Scotch bonnets*, found in meadows, where it grows in circles known as "fairy rings;" pileus seldom exceeds an inch in diameter; stem solid, tough, and fibrous, with a boss or *umbo* in the centre, of a light brown colour; the flesh white, and of a pleasant odour.

Agaricus odorus, *Anise Mushroom*, or *Sweet-scented Agaric;* pileus slightly convex, about three inches broad, and with pale gills; the scent like that of anise; stem strong, fleshy, but not very tall.

Agaricus formosus, or *Smoky Mushroom;* so called from the colour of the upper surface of the pileus; the stalk and gills of a pale yellow; grows in fir woods.

Agaricus primulus, or *Mousseron*, grows in woods and pastures, where the soil is sandy; pileus convex, yellow, about three or four inches in diameter; gills change from white to flesh colour.

How many Vegetable Species exist over the whole surface of the Globe?

I will not do my readers the injustice to suppose that they are unacquainted with the writings of our greatest English poetess, Elizabeth Barrett Browning. They will not fail to have been attracted by the prodigal genius, the superabundant power, the exquisite imagery, the profound spirit of tenderness, the high, pure thoughts, which render almost every page such delightful reading. Successful as she was, however, in giving expression to the most subtle emotions and the intensest feeling, I think she was even happier in her descriptions of scenery. These are invariably aglow with life and colour, and have all the fidelity of Creswick with the imaginative insight of Turner. Turning over her "Aurora Leigh," the other day, I lighted on the following beautiful picture :—

> "I flattered all the beauteous country round,
> As poets use—the skies, the clouds, the fields,
> The happy violets, hiding from the roads
> The primroses run down to, carrying gold—
> The tangled hedgerows, where the rows push out
> Their tolerant horns and patient churning mouths
> 'Twixt dripping ash-boughs—hedgerows all alive
> With birds, and gnats, and large white butterflies,
> Which look as if the May-flower had caught life
> And palpitated forth upon the wind :
> Hills, vales, woods, netted in a silver mist;
> Farms, granges, doubled up among the hills,

And cattle grazing in the watered vales,
And cottage chimneys smoking from the woods,
And cottage gardens smelling everywhere,
Confused with smell of orchards."

FIG. 75.—"And cattle grazing in the watered vales."

As I read this fine passage, the thought occurred to me, How many thousands there are who, in such a scene as it so vividly depicts, would see no beauty whatever, whose heart would not respond to it, whose sympathies would not be aroused by all its variety of outline and all its rich magnificence of colour! Yet not in so wide a landscape alone, but in the smallest nook,—in the little clump of elms by the side of the stream, in yonder grassy knoll rising straight up from the old churchyard, in the quiet angle of the green pasture-meadows, —there is a whole world of wonder and beauty for him who has eyes to see and a heart to feel! Look at the flowery bank which runs along the side of an English lane. Is it not crowded with objects of the rarest and purest interest? Count the many varieties of grasses which clothe it so abundantly, count the many species of flowers and herbs which adorn it with a grace beyond all human skill, and acknowledge that in itself it might supply the inquirer with matter for years of study and meditation.

Pursuing this train of thought, I was led to think of the number of genera and species into which the plant world is divided,—a remarkable proof, not only of the power and wisdom, but of the goodness of the Creator, of His desire to furnish man with inexhaustible sources of pleasure and entertainment; and finally, to put to myself the question, How many vegetable species exist over the whole surface of the globe? If this corner of a leafy English lane is so rich in variety, what must be the case with "the wide, wide world?"

I was now brought to see that a question so difficult could, like so many others, be usefully approached only by its *inferior limit;* in other words, that in the actual condition of botanical science, we can but affirm the number which certainly *exceeds* the sum of the vegetable species scattered over the surface of our earth. To determine this total with mathematical accuracy, we should need to have explored the terrestrial crust, liquid and solid, land and water, from the bed of ocean to the line of perpetual snow, and from the equator to the poles. And as yet we are very far from having obtained so complete a possession of the planet which has been assigned as a dwelling-place to our poor humanity,—alas, more presumptuous than powerful!

The number of plants mentioned by Theophrastus, Dioscorides, and Pliny, whom we take to be the representatives of ancient botany, does not exceed five hundred species. How very few, compared with the presumable total! The Middle Ages added scarcely anything to the botanical researches of antiquity. It is only since the discovery of America that we have seen the domain of Flora extending itself in unexpected proportions. But we must come down to the epoch of Linnæus (the middle of the eighteenth century) before we can obtain an accurate list of species, scientifically classified. Murray's edition of the "*Specilegium*" of Linnæus contains two thousand and forty-two species, including the Cryptogams. Wildmore, in another edition of the same great work, raised the total to twenty thousand. And this was the point at which our botanists had arrived when the nineteenth century opened.

But it was not long before they perceived that all these

estimates, large as they seemed, fell immeasurably short of the reality. In attempting to distribute the different species among the then known regions of the globe, Alexander von Humboldt arrived at a total of forty-four thousand species, Phanerogams and Cryptogams included. De Candolle extended the estimate to upwards of fifty-six thousand.

Let us divide, in fancy, the earth into two parts,—one which has been visited by travellers, and one which still remains to be explored. Can you determine which would present the larger area? The latter.

Thus we possess but a very imperfect knowledge of the luxuriant, the glowing vegetation of the tropical and subtropical regions of the New World, in spite of the labours of Bates, Agassiz, Wallace, and others. To the north of the equator, we know very little of the flora of Yucatan, Guatemala, Nicaragua, the isthmus of Panama, the Chaco of Antioquius, the province of Los Pastos. We are not much better acquainted with the vegetation of the countries south of the equator. What do we know of the manifold species flourishing in Paraguay, in the province of the Missions, in the immense wooded region between the Ucayali, the Rio de la Madeira, and the Tocantin, three affluents of the mighty river Amazon? We know scarcely anything.

Our ignorance increases if from America we pass to Africa. Nearly the whole interior of this continent, from 15° N. latitude to 20° S. latitude, is, botanically speaking, a blank to us. The same is the case with the greater portion of Central Asia. The floras of the south and south-east of

Arabia are still sealed letters,—treasuries to which we have not found the key. As much may be said of the floras of the countries situated between the Thian-Schan, the Kuen-lung, and the Himalaya, as well as of the floras of western China, and most of the trans-Gangetic countries. We know still less of the vegetation of the interior of Madagascar, Borneo, New Guinea, and the greater part of Australia. To conclude: we are probably not acquainted with more than one-fifth of the vegetable species which cover the surface of our globe.

There are regions, moreover, which we imagine will always lie outside of our sphere of investigation; such, for instance, are the Polar regions, properly so called. Undoubtedly, it is open to us to conjecture that the Poles—those two extremities of our axis of planetary rotation—are not the home of any form of life. But this is only a conjecture; we are even without an analogy for it; since we have found, as shown in an earlier chapter of the present volume, living beings, plants, and animals, among the snows of our loftiest mountains. Moreover, might not the auroras, whose maximum of intensity occurs exactly at the poles, render life possible in regions where we at present suppose it to be *im*possible? Conjecture for conjecture,—acknowledge that we here touch in both cases upon an element completely beyond our human power.

These, then, are the reasons why, at present, we can only venture upon defining the *lower* limit, the *restricted* number, above which we are unable to fix the total of vegetable species living on the surface of our planet.

The method to be adopted has been indicated by Alexander von Humboldt in his "Ansichten der Natur" ("Pictures of Nature"). His method consists in the comparison of the vegetable families whose numerical relations are *known*, with the number of species contained in our herbariums, or cultivated in our botanical gardens.

But here, at the outset, a difficulty confronts us. Does any relation exist between the classification of plants by natural families and by their geographical distribution?

To group plants according to their analogies of structure, we study them from an abstract point of view, and without any regard to the medium in which they flourish. The question grows complicated if we also take into consideration their characteristic conditions and their distribution over the terrestrial surface. Families are then split asunder, and the importance of our scientific classifications disappears.

The gathering together of a small number of species, represented by innumerable individuals, confined within the same area, may suffice to communicate to a landscape its characteristic physiognomy: as, for example, is the case with the Asiatic steppes, the *landes* of Brittany, the moors of Scotland, the palm-groves and the clumps of Cactaceæ of tropical America. By the side of species which impress us by their *mass*—that is, by the frequent reproduction of the same individuals at an infinitesimal distance from one another—are placed those much more numerous species which are everywhere very thinly sown.

But do the plants themselves follow, from the equator to the poles, the same law of *decrease* as obtains from the base to the summit of the loftiest equatorial mountains?

Under identical isothermal lines, is the ratio of families known and identified to the probable aggregate of Phanerogams the same, in the temperate zone, on either side of the equator?

What are the vegetable families which preponderate at the two extremes, represented by the torrid and the frigid zones?

Under the same geographical latitude, or between the same isothermal lines, are the Synantheræ, the Gramineæ, the Leguminosæ, the Labiatæ, the Cruciferæ, the Umbelliferæ, more numerous in the Old than in the New World?

What families, either through their mass of individuals or their number of species, take precedence of the other Phanerogams?

How many species of one and the same family belong to any particular country?

What groups or families are characteristic of each zone?

Is the present classification of genera and species in all respects what could be desired?

These are questions that require to be considered, and to some of them we shall presently attempt replies.

Herbariums, though their classification is too frequently imperfect, may furnish us with data of great utility. The great herbarium of Benjamin Delessert was estimated, after his death, to contain 86,000 species,—a total not widely differing from that which Lindley, in 1835, estimated as the probable aggregate of the vegetable species of the world.

Great in importance are botanic gardens. Loudon, in his *Hortus Britannicus* (ed. 1832), places at 22,660 the number of Phanerogams cultivated in the gardens of the Bristol amateur botanists. With this number we must not confound the living species exhibited, in other counties, in gardens designed for the instruction of students, nor the grand total reared for a similar purpose at Kew. Kunth's enumeration, in 1861, of the plants at the Botanic Gardens at Berlin, one of the richest in Europe, amounted to upwards of 14,000 species, including 375 heaths. Among the Phanerogams were 1600 Synantheræ, 1150 Leguminosæ, 428 Labiatæ, 370 Umbelliferæ, 460 Orchidaceæ, 60 Palmaceæ, 600 Gramineæ and Cyperaceæ, &c. By comparing these data with the number of species described in the works of De Candolle, Walpers, Bentham, Lindley, Kunth, and others, we find that in the Berlin gardens are cultivated only one-seventh of the known species of the Synantheræ, one-eighth of the Leguminosæ, one-ninth of the Gramineæ, and about one-fiftieth of the smaller families, such as the Labiatæ and Umbelliferæ.

Now, if we admit that, on the one hand, the number of phanerogamous species cultivated in all the great gardens of Europe is about 30,000, and, on the other, that the cultivated Phanerogams form about one-eighth part of the species described in books and preserved in herbariums, we obtain a total of 24,000 species.

But the Cryptogams, or Agams, such as heaths, mosses, lichens, mushrooms, fungi, mould, and the like, of which our knowledge, as yet, is very imperfect, are probably much more numerous in species than the Phanerogams; for these vegetables, mostly microscopical, develop themselves wherever life can manifest itself—on the barren and denuded rocks, as well as in the air and in the depths of the ocean. If we suppose that they exceed only by 2000 the estimated number of Phanerogams, we shall obtain a total of just half-a-million!

Such, in our opinion, is the number which approximatively represents the lower limit of the aggregate of vegetable species (phanerogamous and cryptogamous) inhabiting our planet. The innumerable individuals of this half-million of species are born, and live, and reproduce their kind, and die, like the twelve hundred millions of individuals of our solitary human species. The former, it is true, remain fixed to the soil which has witnessed their birth, while the latter wander, more or less freely over the terrestrial surface. Do not animals enjoy the same privilege of locomotion? Undoubtedly. But men boast of the reason and the conscience with which they are endowed. Agreed. But with the exception of a small number—the infinite minority of progress—to what advantage have men employed the reason and the conscience of which they boast?

But this is a digression. We proceed to place before the reader a few final data in illustration of the subject we have been considering—the number of existing vegetable species.

The following is an estimate of the known species of plants on the globe at different dates:—

	Phanerogams.	Cryptogams.	Total.
According to Linnæus, 1753,	5,323	615	5,938
Pusoon, 1807,	19,949	6,000	25,949
Stendel, 1824,	39,684	10,765	50,649
Stendel, 1841,	78,000	13,000	91,000
Stendel, 1844,	80,000	15,000	95,000

The advance made of late years in the knowledge of existing species will be apparent from a consideration of Lindley's estimate in 1846:—

	Genera.	Species.
Thallogens,	939	8,394
Acrogens,	310	4,086
Rhizogens,	21	53
Endogens,	1,420	13,684
Dictyogens,	17	268
Gymnogens,	37	210
Exogens,	6,191	16,225
Total,	8,935	92,920

According to Hinds, the following families are almost entirely restricted to particular divisions of the globe:—

To Europe—Globulariaceæ, Ceratophyllaceæ.

To Asia—Dipterocarpaceæ, Aquilariaceæ, Camelliaceæ, Moringaceæ, Stilaginaceæ.

To Africa—Bruniaceæ, Brexiaceæ, Belvisiaceæ, Penæaceæ.

To North America—Sarraceniaceæ.

To South America—Rhizobolaceæ, Gillesiaceæ, Calyceraceæ, Vochysiaceæ, Simarubaceæ, Monimiaceæ, Humiriaceæ, Papayaceæ, Gesneraceæ, Lacistemaceæ.

To Australasia—Goodeniaceæ, Epacridaceæ, Stackhousiaceæ, Brunoniaceæ, Tremandraceæ.

[A group of plants occurring only in one of the six great divisions of the world is called *monomic*, (from μονος *one*, and νομος, a *region*).

A group common to two divisions is *dinomic;* to three, *trinomic;* to four, *quatrinomic;* to all the divisions, *polynomic.*]

NATURAL FAMILIES PREDOMINANT IN THE NORTHERN HEMISPHERE.

Aceraceæ,	Cistaceæ,	Magnoliaceæ,
Alismaceæ,	Coniferæ,	Onagraceæ,
Amentaceæ,	Cruciferæ,	Orobanchaceæ,
Artocarpeæ	Dipsacaceæ,	Papaveraceæ,
(*fam.* Articaceæ),	Elæagnaceæ,	Ranunculaceæ,
Aurantiaceæ,	Fumariaceæ,	Residaceæ,
Berberaceæ,	Grossulariaceæ,	Rosaceæ,
Boraginaceæ,	Hamamelidaceæ,	Rutaceæ,
Campanulaceæ,	Hippocastaneæ	Saxifragaceæ,
Caprifoliaceæ,	(*fam.* Sapindaceæ),	Umbelliferæ,
Caryophyllaceæ,	Hypericaceæ,	Vacciniaceæ.

NATURAL FAMILIES PREDOMINANT IN THE SOUTHERN HEMISPHERE.

Amaryllidaceæ,	Heliotropeæ	Pittosporaceæ,
Atherospermaceæ,	(*fam.* Ehretiaceæ),	Polygalaceæ,
Cactaceæ,	Iridaceæ,	Proteaceæ,
Capparidaceæ,	Malpighiaceæ,	Restiaceæ,
Crassulaceæ,	Melastomaceæ,	Scævoleæ
Dilleniaceæ,	Mesembryaceæ,	(*fam.* Goodeniaceæ),
Diosmeæ	Myoporineæ	Spigeleæ
(*fam.* Rutaceæ),	(*fam.* Verbenaceæ),	(*fam.* Loganiaceæ),
Geraniaceæ,	Myrtaceæ,	Stylidiaceæ.
Hæmodoraceæ,	Oxalidaceæ.	

The Harvest Bug.

"I very much wish," said my friend T. to me one day, "to buy a small estate in the vicinity of —— Forest. If there should be one to sell, pray let me know of it."

It was not long before an opportunity arose for my friend to satisfy his desire. But after I had made him acquainted with it, he declared himself no longer willing to purchase a property in a district where, as he had learned, one was devoured by *red beasts* all through the finest months of the year. What a frightful neighbourhood to live in, where you were forbidden to walk in your garden under pain of catching an itch in your legs!

Unquestionably, it is only too true that the cultivated ground, whether on the northern or the southern slope of the forest, is infested, from the beginning of summer to the beginning of winter, by Lilliputian horrors, like so many tiny red points, which cling obstinately to the skin, and there deposit, under the epidermis, their microscopic brood. Once planted there, the *rougets*, as the French call them, or *harvest bugs*, as we English call them, effect considerable mischief; and if, to relieve one's self, one indulges in "a scratch," the cutaneous surface is quickly covered by small blisters, which on a cursory examination might be taken for a skin affection not generally named in polite hearing.

But one does not perceive the galleries excavated by these annoying insects, positive tunnels or covered ways, through which they proceed to pour forth elsewhere the superfluity of their numerous progeny. Less prolific than the *Acari*, which create upon the skin immense patches of irritation, the harvest bugs confine themselves to a few circumscribed localities: their favourite choice being the legs, the arms, and the corners of the eyes, especially among young children. They are not above domestic animals; cats and dogs frequently suffer from them,—not, indeed, over the whole surface of the body, for they are not so wandering as the *Acari*,—but particularly inside the shell of the ear.

At the first glance you would scarcely believe that those red points, apparently immovable, could be living beings,—could be animals belonging to an order of some importance.

Let us attempt to isolate one of the animalcules with the point of a pin: it is not an easy thing to do, because they usually adhere to the epidermis in clusters of three or four individuals. There, now we have succeeded, and here is one before us: it is only the fifth of a millimetre in diameter, which is, for most people, the very last limit of the visual function (see the small white line in Fig 76, *a*). And, in truth, it *would* be imperceptible to the eye but for its bright red colour. To study it carefully, of course, you must make use of a very strong lens, or, rather, of a microscope. (See Fig. 76, *b*.)

FIG. 76.—*a Leptus Autumnalis* (nat. size). *b* Ditto (mag.)

To this tiny animal has been given the name of *Leptus autumnalis*; the first, on account of its extreme delicacy; the second, because it is visible up to the end of autumn.

When examined through a microscope, it produces on the spectator the impression of a spider; but, like all other insects, it has only six legs.

Our naturalists, however, have found some difficulty in classifying it; and by way of cutting the Gordian knot of their embarrassments, some have placed it in a separate family of *Microphthiræ* (literally "little lice"), which is made to include all Arachnidæ with six legs. Others, who regard the wheat worm as an insect, rank it among the parasitical Apteræ.

In effect, it has all the characters of the parasitical insect— its protracted head, distinct from the rest of its body, is sometimes thrust forward in quest of its food, sometimes drawn back or concealed, to protect it from danger. Intended to suck rather than to knead or bruise, it has a sucker, but no mandibles. The head is without antennæ, and its palpi are very short, barely visible, and of a conical form. The body is ovular and very soft (whence the Grecian name *leptus*, λεπτός, signifying "soft"). The anterior part, corresponding to the thorax, is broader than it is long, and is marked underneath, on each side of the central line, by a black point: these two points, symmetrically placed, appear to represent the eyes.

The posterior portion, corresponding to the abdomen, is longer than it is broad, and covered with hairs. Each leg consists of six joints, easily distinguished by the hairs inserted at each articulation; and each terminates in a couple of strong crooked *claws*, which enable the animal to obtain a firm hold on the skin.

Thus, then, to judge from the aggregate of its characters, the harvest bug, *Leptus autumnalis*, belongs to the class Arachnidæ, while the number of its feet places it in the class Insects. But this is a detail which causes little annoyance to a person being devoured by the "red beasts," and only anxious to rid himself of them.

But if such be his desire, let me tell him that the best remedies are bathing the afflicted part with lotions of vinegar, or rubbing it with sulphur ointment.

I have been asked whether certain tiny parasites, such as the *Ocypete rubra*,—which is also red, and has six feet like the *Leptus autumnalis*, but which, instead of attacking man and his domestic companions, attaches itself to flies,—I have been asked whether these insectiform Arachnidæ may not be species of larva not yet arrived at their matured condition.

For my part, I must acknowledge that, whether the *Ocypete rubra* is or is not the transitory state of a more perfect animal, I do not know. But I am sure that the *Leptus autumnalis* lives and dies on the skin where it has selected its dwelling-place,— a *living* dwelling-place.

It is impossible to be too circumspect in the determination of certain genera and species, whose different phases of existence are little known, and which seem, so far as their characteristics are concerned, to participate of several orders or classes of articulated animals. The errors which have been committed in this respect ought at least to teach us caution.

Thus, the red, oval, six-legged animalcules, whose mobile heads are furnished with a proboscis shaped like an angular beak, and whose two palpi are large and semi-transparent,— the singular animalcules which, in June or July, are hatched in the spongy stems of certain aquatic vegetables,—notably the *Potamogeton natans,*—have been described* as forming a peculiar genus of Arachnidæ, the genus *Achlysia;* and this genus, created by Audouin, was ranked along with the *Leptus* and *Ocypete* in the family of Microphthiræ. Yet nothing is less exact. These Achlysia are simply the larvæ of a kind of *Hydrachna* or water-acarus. To be convinced of this, you have but to watch their development. At first very small and pear-shaped, these larvæ, deprived, like all larvæ, of the reproductive organs, rapidly increase in size. At the end of a few weeks you will see them adhering to a leaf of potamo- geton; they thrust their proboscis into the stem, and cling to it with their palpi. Little by little, the legs, the proboscis, and the palpi, are drawn back towards the body, abandoning the skin which has hitherto formed for each of these organs a kind of horny sheath. From the larva state, the animal passes into that of the nymph. But this nymph continues to feed and enlarge; proboscis, legs, and palpi grow thinner and harder; claws, ciliæ, and hairs are developed; and, finally, through a fissure in the skin emerges the perfect animal, red as wine, with eight feet, and about two millimetres in length.

* "Annales des Sciences Naturelles," i. 18.

This animal, placed in the family of the *Hydrachnellæ*, has been described by De Geer under the name of *Acarus aquaticus globosus*, and by Dugès under that of *Hydrachna globosa*, on account of its globular form.

The Cheese-Mite.

From the crust of a dry old cheese,—such a kind of cheese as a *bon-vivant* likes with a glass of "good old ale,"—a very fine powder often crumbles off, like the dust made by wood-eating worms.

Examine this powder with your lens, or if you have good eyes, you may make use of *them*. You will quickly detect something moving in it, and by degrees you will see that this movement pervades the whole mass; that there is a general stir and commotion in all directions.

But you find it impossible to distinguish clearly the form of the animals which are thus agitated. You are certain, however, that they are not maggots, for *they* affect moist cheeses; besides, they are visible enough to everybody, and at need can make themselves *felt* upon your hands, and even upon your face, for they have a faculty of launching themselves to a distance by a little manœuvre familiar enough to serpents: bringing the head round towards the tail, they curve themselves like the spring of a watch, then abruptly uncoiling themselves with the help of some solid *appui*, they fling forth into the air, and are thus launched to very considerable distances. It is a curious species of locomotion, not unworthy the attention of the mechanician.

To clear up the mystery of a movement whose cause is not apparent at the first glance, let us sprinkle with this impalpable débris, — with this kind of sawdust, or *cheese-dust*,—a little strip of glass, and place it beneath the focus of a microscope.

Ah! you exclaim, what a frightful creature! These long sharp ciliæ seem to be so many lancets covering the whole body, and especially the legs; its head, like that of the harvest-bug, protrudes and recedes under a transparent carapace; thus communicating to the animal something of the aspect of a turtle. In all other respects its form exactly resembles the harvest-bug; only its body is more elongated towards the anterior extremity than that of the latter. While the harvest-bug makes us think of a spider, the body of the Acarus has a greater likeness to an insect's. (Fig. 77.) Yet the Acarus has eight legs, like a spider, and the harvest-bug six, like an insect. Attempt, then, to establish your absolute rules!

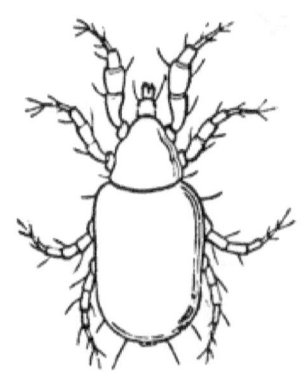

FIG. 77.—The *Acarus domesticus.*

Let us continue our observation of this cheese-worm. The well-defined thorax forms nearly one-third of the fore-part of the body, which is of a shining whitish-red or reddish-white. The proboscis, shaped like a conical tube, is armed with two projecting mandibles, which, like true pincers, can be brought close together, or moved wide apart, thrust forward singly or simultaneously. Our animal, which a small lens makes very distinct, has been more than once confounded with the *Sarcoptes scabiei*.

Let us resume. Our cheese-dust, which to all appearance walks alone, encloses legions of mites; the old you may detect by their eight feet, the young by having six. The germs, or eggs, whence they spring, are found mixed among the excrements of the living, and the débris of the dead.

It is in this way that a crust of cheese offers us a true, a vivid image of the terrestrial crust. So may we learn to compare small things with great.

How many Animal Species are there Distributed over the Surface of the Globe?

In the present condition of scientific knowledge, no satisfactory answer can be given to this important and most interesting question.

The truth is, that what we may call Geographical Zoology is as yet in its very infancy. The few works which have been published on the subject have been published within the last eighty or ninety years; and they embrace only the vertebrate animals, notably the mammals, birds, and reptiles, or amphibia. We shall attempt to place before the reader an outline of the results that have so far been obtained.

Of all the Vertebrata, we are best acquainted with the mammals. And yet our zoologists differ very widely in respect to the number of their species, though the calculations have been made at very short intervals. For instance, in 1829, Minding computed that the globe contained 1230 species of mammals. In 1832, Charles Bonaparte reduced the total to 1149. Oken estimates it at 1500; and this last figure would seem to be the most probable.

Nothing is more curious than the distribution of these 1500 species of mammals, according to the different regions and climates of the globe.

Man, according to the best-considered data of science, forms a single family, a single genus, a single species. He alone possesses the power of adapting himself to every climate, and of taking possession of countries the most widely opposite in character. We find him among the snows of the North Pole; we find him under the blazing sun of the Tropics. We find him in the palm-fringed islands of Southern Seas, and in the barren burning waste of the inhospitable Sahara. Considered as an animal who feeds and reproduces himself, he forms alone the order of *Bimana;* so named in opposition to the *Quadrumana,* or apes, who make use of their fore-feet as we do of our two hands. Deprive man of his progressive and transmissible intellect—of those mysterious powers which we call the mind and the soul—and he would become at once the most useless and the most wretched member of the animal world.

The warm regions of the old and new continents are the true home and haunt of the apes. They are not sufficiently developed to be able to frequent the temperate or frigid zone. In our European menageries the specimens nearly all die of consumption. The Quadrumana form about one-fourteenth of the whole number of species of Mammalia.

The *Carnivora*, characterised by the development of their canine teeth, are spread over the whole globe. They are found in greater numbers in the torrid, however, than in the frigid zone. Their species compose at least one-third of the Mammalia.

The *Rodentia*, characterised by the development of the incisors, are wanting in Polynesia, and are rare in Australia. They are found in their maximum number in the torrid zone. Like the Carnivora, they form about one-third of the Mammalia.

The *Ruminantia*, remarkable for the development of their digestive apparatus, are distributed into 165 species, representing something less than one-ninth of the Mammalia. Africa, of all the continents, is richest in the Ruminants.

The *Marsupialia*, so strangely distinguished by the membranous pouch in which they enclose their young, belong to America, and especially Australia. At present about 123 species are known, or a little more than one-thirteenth of the Mammalia.

The *Edentata*, so named on account of their incomplete dentition, inhabit the tropical regions of the Old and New

World. They are distributed into 32 species, 19 of which belong to America. The Edentata, therefore, do not form more than one-fiftieth of the Mammalia.

The *Pachydermata*, which owe their name to the thickness of their skin or hide, almost exclusively belong to the Old World. None are found in Australia. The number of their species is 38, of which 5 only belong to Southern and Central America. The Pachyderms form, therefore, nearly one-thirty-seventh of the Mammalia.

The *Cetaceæ*,—which the naturalists of antiquity ranked among the fishes, though the females bear their young alive, and are furnished with a mammary apparatus,—chiefly frequent the Northern waters, but some of their species are found in the South Pacific. They represent, it may be assumed, about a one-hundredth part of the Mammalia.

The *Birds*, by their feather-clad bodies, and by the transformation of their two fore-limbs into wings, form the best-characterised class in the whole animal kingdom. But naturalists can no more agree as to the number of their species than as to the number of species composing the Mammalia. Some, taking as a foundation the rich ornithological collection in the Berlin Museum, allow for 6000 species being distributed over the surface of the globe; others, like Lessen, increase the total to 6266; while Dr Gray, no mean authority, raises it to at least 8000.

The majority of the *Raptores*, or birds of prey (vulture, falcon, eagle), as well as nearly all the *Waders* (stork, crane, heron), and *Palmipedes* (duck, goose, water-hen), are cosmopolitan birds. The other orders, such as the *Scansores* (parrot, parroquet, magpie), the *Passeres* (comprising nearly all the singing birds), and the *Gallinaceæ* (pheasant, pintado), prefer, as a general rule, the warm temperate regions. They are not found in the extreme north, nor in the equatorial climes, except in limited numbers.

SUMMARY OF THE MAMMALIA.

Assumed total, 1600 *species.*

Bimana	form	1	species.	Marsupialia	form	123	species.
Quadrumana	,,	105 (?)	,,	Edentata	,,	152	,,
Carnivora	,,	510 (?)	,,	Pachydermata	,,	38	,,
Rodentia	,,	508 (?)	,,	Cetaceæ	,,	18 (?)	,,
Ruminantia	,,	165	,,			1600	

[Of course, the foregoing is but an approximative estimate, but it will provide the reader with a tolerably accurate notion of the proportion borne by the different classes of Mammalia.]

About 5000 species of birds have been classified. By Cuvier's system they are divided into six orders:—

1. *Raptores*, or birds of prey.
2. *Passerine birds*, now generally called *Insessores*, or Perching-birds.
3. *Scansores*, or Climbing, frequently called *Zygodactyli*, or *Zygodactylous* birds.
4. *Gallinaceæ*, now more frequently known as *Rasores*.

5. *Grallatores*, Waders, or Stilt birds.
6. *Palmipedes*, or Web-footed birds, now more generally recognised as *Natatores*, or Swimmers.

It has been proposed to separate the *Brevipennes*, or short-winged birds, from the Grallatores, and erect them into a separate order.

The *Reptiles*, of which the majority possess the faculty of living upon land and in water,—whence their name of Amphibia,—never pass beyond the limits of warm and temperate climates: their blood, which has the same temperature as the medium wherein they live—whence their name of "cold-blooded animals"—does not circulate where the mean annual temperature descends below freezing-point. Yet frogs and salamanders have been met with in Greenland, and on the banks of the Mackenzie River, in North America, under 67° latitude.

Linnæus was not acquainted with more than 215 species of *Amphibia*, divided into four orders:—the Chelonians, or tortoises; the Saurians (as the lizard and crocodile); the Ophidians (serpents); and the Batrachians (frogs). In 1789, Lacépède raised the total to 303; in 1820, Merrem estimated it at 677. At present, the number of species of Reptilia classified and described amounts to 2000, and the four orders into which they are distributed are—

1. *Ophidia*, or Serpents.
2. *Sauria*, or Lizards.
3. *Loricata*, or Crocodiles.
4. *Chelonia*, or Tortoises.

According to Sching, there are 7 tortoises, 33 serpents, and 35 lizards.

Fishes are the least known of those superior animals whose skeleton and vertebral column are situated in the interior of the body, and which are thence named *Vertebrata*. The richest collections, such as those of the British Museum, and those of the Museum of Natural History in Paris, which contain about 3000 species, do not represent probably more than a fourth of the existing total, including fresh-water and salt-water fish. How many rivers and streams in both hemispheres still remain to be explored! How far we are from a knowledge of the fishes which people the different strata of the great ocean.

Agassiz divides this great class of vertebrated animals into the four orders of *Cycloid, Ctenoid, Placoid, and Ganoid*, according to the character of their scales. Cuvier, into *Osseous* fishes (with true bones), and *Cartilaginous*; subdividing the former into *Acanthopterygii* and *Malacopterygii*.

The difficulty of the problem we are here considering increases when we come to the inferior animals. Who would pretend to determine the number of species of *Mollusca* which inhabit the earth, the fresh waters and the salt? This much is certain, that it cannot be less than that of the Vertebrates.

In the vast aggregate of the *Articulata*, the inquirer finds himself utterly astray and bewildered. This great division is not divided into those which have, and those which have not, articulated members.

The first subdivision includes *Insects, Arachnida, Crustacea,* and *Myriapoda*; the second, *Annelida* and *Entozoa.*

Some naturalists, be it said, rank the *Cirrhopoda* as intermediate between the two; others place them among the Mollusca. Others, again, include the *Rotifera* in the second sub-division.

We shall in this place confine our remarks to the *Insects.* According to the most distinguished entomologists, the average number of species at present, described or not described, and preserved in entomological collections, is between 150,000 and 170,000.

This estimate is obviously below the truth. Take only the Coleoptera, which forms but one, though, it is true, the most numerous order of insects. Thirty years ago the most complete collections contained about 7000 species. In 1850, the museum at Berlin, according to Alexander von Humboldt, contained nearly 32,000. We would here call the reader's attention to the just remarks of the author of the "Natural History of the Coleoptera," an entomologist of great authority, whom a long residence in America had peculiarly qualified to pronounce an opinion on the subject before us:—

"If we remember," says the Count de Castelnau, "that there are immense regions in Asia and the two Americas of which we do not possess a single coleoptera; if we reflect that the interior of the vast continent of New Holland is, from this standpoint, entirely unknown, and that most of the archipelagoes of the great ocean have never been entomologically explored, we may conclude, without any fear of mistake, that the number of existing coleopteras exceeds *one hundred thousand*. However frightful this number may appear, it will seem less so if we examine only the species discovered in the neighbourhood of Paris, within a radius of twelve to fifteen leagues; and we do not hesitate to say, that in a few years the Parisian fauna alone will present material for a considerable work, which shall not treat of less than 3000 to 4000 species of Coleoptera." *

If we admit that the other orders of insects, the Lepidoptera, the Hemiptera, the Hymenoptera, the Neuroptera, the Orthoptera, the Diptera, the Strepsiptera, comprise, taken altogether, at least the same number of species as the Coleoptera alone, we shall gain, for the class of insects, a total of 200,000. And we shall certainly keep within the truth if we assign the same number of species to the Annelida, the Crustacea, the Arachnida, the Myriapoda, and the Monomorpha, to which, with some modification, we may apply the remarks already called forth by the Coleoptera.

Let us recapitulate. The four classes of *Vertebrate Animals* include approximatively:—

1,600 species	. .	Mammals.
5,000 ,,	. .	Birds.
2,000 ,,	. .	Reptiles.
12,000 ,,	. .	Fishes.
20,600		

---

* Count de Castelnau, "Histoire Naturelle des Coléoptères," i. 7.

If we add to these 20,600 species of Vertebrate Animals, 200,000 species of Articulata, and 22,000 Mollusca (a minimum), we shall have a total of 242,600.

But to complete the grand whole of beings "who grow, and live, and feel" (the definition of animals laid down by Linnæus), we must add the Intestinal Worms, the Echinodermata, the Acalephæ (or Sea-nettles), and the Polypes. The history of these singular creatures, which apparently form the transition between the animal and vegetable kingdom, and have thence been designated *Zoophytes*, leaves much, very much, to be desired before it will be possible to indicate, even approximatively, the number of their species.

And, finally, what shall we say of the Infusoria? These microscopic forms of life seem, by their extreme multiplicity, to animate all nature. It is in studying these that the inquirer needs to be constantly on his guard, that he may not mistake transitory conditions—or larvæ—for actual species, and it behoves him to understand thoroughly the difficult delimitation of specific characters. It would be far easier to ascertain the exact number of human beings who at present people the terrestrial surface, than to fix the total of the species of Infusoria now in existence; assuredly it exceeds 250,000. What an infinite variety of design is here! What a picture it presents of the inexhaustibility of the Creative Mind!

Add, then,—let us say, in conclusion,—to this last great total the aggregate of the Vertebrates, the Articulates, and the Molluscs, and for our grand whole we have a minimum of *half a million* of ANIMAL SPECIES! This is the very figure, observe, at which we arrived as representing the lowest limit of the totality of VEGETABLE SPECIES, living and moving, flourishing, and dying, and reproducing, on the surface of the globe.

We leave the reader to meditate—as meditate he surely must—on the sublime thoughts, the overpowering ideas of Power and Wisdom which these considerations suggest.

WHAT IS CHLOROPHYLL?

We are drawing towards the close of autumn; we shall soon be in sight of the "melancholy days of the year;" when, for a while, the "voice of the turtle" will cease in the leafless groves, and the banks and braes will be sadly bare of their floral garniture. As yet, however, the trees retain their glorious vesture, though streaked and varied with the gorgeous colours of decay; and in the sheltered corners of the woods, on the sunny southern slope of the grassy hill, and beneath the covert of the still fragrant hedgerow, many a blossom appeals to our souls with its promptings of sweet images and tender fancies. The arum still raises its clusters of deep-scarlet berries, and spreads its spotted leaf—

"Armed with keen tortures for the unwary tongue;"

the blue-bells hang their delicate cups among the thick herbage; and the wild marigold contrasts its yellow splendour with all this crimson and azure magnificence. The daisy, too, has not forsaken us—sweet shield of silver, embossed

Fig 78.—"As yet, the trees retain their glorious vesture."

with gold!—but brightens still the pleasant meadow and the sloping bank.

> "The rose has but a summer reign,
> The daisy never dies;"

and though it first makes its appearance in the merry spring-time, and is truly a child of the early year, it lingers on to become a precious ornament of our scanty autumn wreaths. Sweet flower of song!—dearer to the poet than even lily or violet!—who does not remember, and remembering feel, all the pathos of the dying exclamation of poor Keats,—" I feel the daisies already growing over me!" They heighten the commonest and cheer the saddest corners of the earth, and are ever ready, in their simple loveliness, to awaken thoughts of grateful tenderness and love—

> "So glad am I when in the daisy's presence,
> That I am fain to do it reverence."

To what do the leaves, now changing their hues so rapidly, and varying through all the tints of purple, brown, and yellow,—to what do they owe their normal colour, the fresh, vivid, beautiful green?

To a substance called *chlorophyll*—(χλωρὸς, green, and φυλλον, a leaf).

Well, what *is* chlorophyll?

The colouring matter of plants, which, accompanied by grains of starch, floats like very minute seeds in the fluid of their cells. In some respects it is analogous to wax; it will not dissolve in water, but is easily affected by ether or alcohol.

Chlorophyll is dependent upon the action of light, if not for its formation, at all events for its development. Keep a plant in a dark room or cellar, and it will become blanched and sickly; the colouring matter dries up, and the white, wan tissue of the leaf is all that survives. The more a plant is exposed to the light, the deeper will be its green. In a shrubbery you may notice that the brown leaves of any particular evergreen or bush, if so situated as to lose the direct action of the sun's rays, will soon change colour. Instead of their natural brightness of tint, they assume a sickly greenish-yellow hue, and are said to be suffering from *chlorosis*. The formation of the chlorophyll is obstructed, or takes place too slowly. Of course, this peculiar condition will frequently arise from bad soil, or a long continuance of damp weather; but it is also the result of a want of light.

It should be observed that young leaves are always of a lighter green than old; simply because the latter have been exposed for a longer time to the light. And so the leaf goes on deepening and deepening in colour, until the sad days of autumn come, and the green gives way to yellow and brown and red, owing to the influence of the changing season on the chlorophyll of the plant.

In reference to this interesting subject,—which deserves to be more closely investigated,—we may place before the reader the results of certain recent experiments.*

MM. Prillieux, Brongniart, and Roze (*Comptes Rendus*, Jan. 3 and 17) have made some important observations on the apparently spontaneous movements of the grains of chlorophyll within the leaves of plants. These had been observed by Böhm to congregate under the direct action of the sun;

* As recorded, in a condensed form, in *The Academy* (Feb. 12, 1870), pp. 131, 132.

Famitzin, confirmed by Borodine, had also recorded very marked movements in the leaves of a moss under the influence of light. This class of plants offer great facilities for these observations, inasmuch as the movements can be observed in them under the microscope without dissection. M. Prillieux kept a moss in the dark for several days, when the cells presented the appearance of a green network, between the meshes of which was a clear transparent ground. All the grains of chlorophyll were applied to the walls which separate the cells from one another; there were none on the upper or under walls which form the surfaces of the leaf. Under the influence of light the grains change their position from the lateral to the superficial walls; under favourable circumstances this change takes place in about a quarter of an hour. On attaining their new position, the grains do not remain absolutely immovable, but continually approach and separate from one another. If again darkened, they leave their new position and return to the lateral walls. Artificial light produces the same effect as daylight. M. Brongniart further observed that this movement of the chlorophyll, under the influence of light, does not consist in the change of position of isolated grains, but of masses of net-work, each containing a certain number of grains. In addition, M. E. Roze states that, besides the grains of chlorophyll which coat the walls of the cell, each cell is lined with a transparent mucous plasma formed of very fine threads, the extremities of which unite together the grains of chlorophyll. This protoplasm exhibits, under a high magnifying power, a very slow motion, and carries the grains of chlorophyll along with it. M. Roze believes, therefore, that the motion is a plasmic one, the protoplasm being the vital and animating part of the cell.

CARNATIONS AND PINKS.

Among the latest flowers of the autumnal garden are those old favourites, the "July-flowers," or *Carnations*, which, because they were "fair and sweet and medicinal," Jeremy Taylor preferred to "the prettiest tulips, that are good for nothing." I remember a time when they were among the best-prized ornaments of our parterres, and very delicious it was to inhale the balmy breath that rose into the warm air of an autumn evening from rich masses of carnations and pinks. The carnations were also called—*sub consule Planco*—in the merry days when I haunted the green lanes of a pretty Devonshire village, carnations, and clove July-flowers or gilliflowers; and an ancient name for the pink was that of sops-in-wine, because they were infused in the wine-cups of our much-drinking ancestors. So Drayton says :—

> "Bring hither the pink and purple columbine,
>   With gilliflowers;
> Bring coronations, and sops-in-wine,
>   Worn of paramours."

The same poet alludes to them under their more modern appellations :—

> "The brave carnation, then, of sweet and sovereign power
> (So of his colour called, although a July flower),
> With the other of his kind, the speckled and the pale ;—
> Then the odoriferous pink, that sends forth such a gale
> Of sweetness, yet in scents as various as in sorts."

The scientific name of this beautiful family of plants, whose rich dyes are not less conspicuous than their Sabæan odours, is *Dianthus*, or "Flower of God." They form a genus of the

FIG. 79 —"When I haunted the green lanes of a Devonshire village."

natural order Carophyllaceæ; the calyx is tubular, and five-toothed; there are five petals, which at the throat of the corolla are lightened (as it were) into a linear "claw." The stamens are double the number of the petals; the capsule is of a cylindrical outline, and one-celled.

I am quite prepared to agree with a sympathetic writer on flowers that, during summer, and far into the autumn months, the greatest beauty of our gardens is the varied tribe of Carnations, while their exquisite, subtle, yet potent aroma is not to be excelled, I think, or, at all events, is not far surpassed, in strength and sweetness, by the much-lauded rose. A carnation seems, to my humble taste, the very embodiment, as it were, of the favourite qualities so insisted upon by Mr Matthew Arnold, "sweetness and light." And even in winter, when its radiant petals have disappeared, there is something graceful to the eye in the long slender leaves of the pink, covered with their sea-green powdery bloom.

The two species commonly grown in gardens are, the garden pink (*Dianthus hortensis*) and the carnation proper (*Dianthus caryophyllus*); both of which are generally referred to one original, the castle-pink, July-flower, or clove-gilliflower. The carnation, as a garden flower, was originally brought into England from Germany, where it has always been a favourite object of cultivation.

There are several hundred varieties of it, which are arranged into three principal divisions: *flakes*, which are diversified by broad stripes of two colours only; *bizarres*, which are of several colours, and very irregularly streaked; and *picotees* (from *piquetté*, spotted), whose flowers are besprinkled with different colours, and their petals fringed or serrated.

In England, the *native* species of pink are five in number, but they are mostly rare, or, when abundant, are found in very limited habitats.

The commonest kind is the little *Deptford Pink* (*Dianthus Armeria*), which sometimes grows in thick clusters among the meadow grass. In shape, its blossom resembles that of the garden pink; in size, it is about equal to that of the sweet william; and its flowers grow in a very similar manner. It is a scentless pink, however, with serrated or notched petals, and its rose-coloured petals curiously besprinkled with tiny spots of white.

A very pretty species is the *Maiden Pink* (*Dianthus deltoides*), which some botanists think to have been the original of our garden favourite; and a kind deserving notice for its large and fragrant flowers is the *Dianthus superbus*.

The maiden pink, I should add, has delicate rose-coloured blossoms, daintily touched with silver, and a white eye encircled by a deep purple ring. It is not unworthy of its fanciful and highly suggestive name.

A rare British variety is the *Clustered Pink*, or *Childing Pink* (*Dianthus prolifer*), which produces its flowers in plentiful clusters, but is only allowed a season's sunshine.

The *India* or *China Pink* (*Dianthus chinensis*) is a native of Eastern Asia, but has now become a frequent denizen in our English gardens.

One wild species, the *Mountain Pink* (*Dianthus cæsius*), it has never been my fortune to gather in its native home. It is described as a large handsome flower, and it loves to breathe

the "difficult air" of the lofty mountain-top. "Never," we are told, "is it found in plain or valley; but it is one of those blossoms whose beauty gladdens the mountaineer, or bids the traveller wonder that so lovely a flower should be blushing on the lone summit, scarcely accessible to his footstep; or cheering a rock, where only the yellow lichen, or the verdant or gray moss, reminds him of vegetation. Such a sight might bid one think of the old motto, which accompanied a wild flower, 'I trust only in Heaven.' How beautiful is it in its loneliness! Scarce an eye meets it but that of the towering bird, as he dashes through the air above it, yet is it as full of lustre as the flowers we daily see and admire. Surely it should arrest the eye and the thoughts of the traveller as certainly as would a monument of human skill on such a spot. Like a lone ruin, it is a page of story, telling not only of the past, but the present, and reminding us of a Being who has reared it there, where it stands a memento of power and goodness."

> "Thanks to the human heart by which we live,
> Thanks to its tenderness, its joys and fears,
> To me the meanest flower that blows can give
> Thoughts that do often lie too deep for tears."
>
> — WORDSWORTH.

Of greater interest, however, because a native species, and more easily attained, is the *Castle Pink*, to which brief reference has already been made. Its perfume is like that of precious spices, and after a shower of rain, the air, for some distance, is actually interpenetrated with it. As its name indicates, it loves to grow upon the shattered walls of

> "Chiefless castles breathing stern farewells;"

and it may be found, unless swept away by barbarous "improvements," adorning the gray old masonry of Sandown Castle, and on ruins in the neighbourhood of Norwich. On

FIG. 80.—"On the time-worn ruins of an ancient minster."

the walls of Rochester's stately keep it grows at a height which defies the spoiler's hand; and on the time-worn ruins of an ancient minster, it shines, in the summer-noon, "with a flush of flowers." It blossoms in July, and there are not, it is said, more than half-a-dozen spots in England where it may be found wild.

To this dainty and beautiful tribe belongs that common but handsome and most fragrant flower, the *Bearded Pink* or *Sweet William* (*Dianthus barbatus*),—a native of central Europe and southern France, with long lanceolate leaves, bearded petals, ornamental bracts, and dense clusters or tufts of crimson or rose-coloured blossoms. It has long been a favourite with the cottager, for it is so hardy that it will grow in any soil, and will flourish even in the odd corner known as the "children's garden." Its popular name, "long, long ago," was "London tuftes;" and it owes its specific appellation of *barbatus*, or "bearded," to the nature of its calyx. That quaintest of quaint old botanists, delightful Gerarde, lavishes encomiums upon its beauty, and pronounces it meet "to deck up the bosoms of the beautiful, and garlands and crowns for pleasure." I suspect that now-a-days it seldom figures in a posy.

To the same order as the Dianthi—that is, to the *Caryophyllaceæ*—belong many wild flowers of lowly growth but abounding interest; as, for example, the corn-cockle, whose lilac-coloured petals, soaring conspicuously among the tall waving corn, have procured for it the right royal appellation of *Agrostemma*, or "the crown of the field." So, too, the numerous species of campion and catchfly (*Silene*), with their singular expanded calices; and those handsome flowers, the white and rosy lychnises, which love to air their charms by the side of running waters. The cottony down upon these plants was wont to be much used for the wicks of lamps.

Then, too, the whole tribe of chickweeds are included in the Caryophyllaceæ. They are spring flowers, with pearly white blossoms, five-petalled, like a five-rayed star, and long slender drooping leaves. Their resemblance to a star has suggested their scientific name, *Stellaria;* and they are truly the "stars of the earth," glistering among the thick herbage with a modest beauty. The *Stellaria media* supplies our song-birds with an abundant and a wholesome provision.

A handsome wild plant of this order is the Soapwort (*Saponaria officinalis*). It is common in Kent, and some of the neighbouring shires, but in many parts of England is never seen. Its full cluster of rose-hued blossoms is rather larger, and more loosely set together, than those of the sweet william; which, however, it much resembles in its leaves, these being opposite to each other, and nearly sheathing or surrounding the stalk at their bases.

"The juice of the soapwort," says Miss Pratt, "is one of those vegetable substances which, by making a lather with water, will cleanse linen, and remove grease as effectually as soap. It grows more generally in the neighbourhood of villages than in any other situation, as if providence had placed it there especially for the service of the cottager; yet it is very little used, either from ignorance of its properties, or because it would require some cultivation to render it sufficiently plentiful for household purposes. It needs the addition of ashes to make it a good soap for washing linen; but it is of much service to the shepherds on the Alps, who wash their flocks, previously to shearing them, with soapsuds made by boiling this plant in water. The large fruit of the horse-chestnut has similar cleansing properties, and may be used by cutting it into small pieces, or scraping it into water. It has even been suggested that if the nuts were reduced to powder, and made into balls, with some unctuous substance, they would answer all the purposes of our manufactured soap; and yet numbers of poor people see these nuts lying decaying in their neighbourhood, and have no idea of making them of any service."

On the Continent, however, the peasantry are wiser, and not only provide themselves with chestnuts for soap, but gather the beech-leaves to stuff their mattresses.

Returning to the Caryophyllaceæ, we may add that some of the plants of this order have poisonous qualities, which are due to the principle called *saponine* existing in many of the species of Saponaria, Silene, Lychnis, and Dianthus.

According to Lindley, the order includes no less than 53 genera, and 1055 species. They inhabit chiefly temperate and cold regions, and are ranked in three sub-orders,— Alsineæ, Sileneæ, and Mollugineæ.

The Eglantine and the Convolvulus.

What plant do our poets mean by the eglantine? What by

the woodbine? Are they one and the same, or are they different?

We cannot answer those questions until we have referred to three or four passages in which they are introduced. And, first, let us take an example from Spenser:—

> " And over him, art striving to compare
> With nature, did an arbour green dispread,
> Framèd of wanton ivy, flowering fair,
> Through which the fragrant eglantine did spread
> His pricking arms, entrailed with roses red,
> Which dainty odours round about them threw,
> And all within with flowers was garnishèd,
> That when wild Zephyrus amongst them blew,
> Did breathe out bounteous smells, and painted colours shew."

And now from Milton :—

> " Through the sweetbriar, or the vine,
> Or the twisted eglantine."

The following is from Sir Walter Scott :—

> " On the hill
> Let the white heath-bell flourish still,
> Cherish the tulip, prune the vine,
> But freely let the woodbine twine,
> And leave untrimmed the eglantine."

From Burns :—

> " The woodbine I will pu' when the evening star is near,
> And the diamond draps of dew shall be her een sae clear."

From Michael Drayton :—

> " The azured harebell next with them they neatly mixt :
> T' allay whose luscious smell they woodbine placed betwixt . . .
> The columbine amongst they sparingly do set,
> And now and then among, of eglantine a spray."

And lastly, from Shakespeare :—

> "And leaf of eglantine, whom not to slander,
> Out-sweetened not thy breath."

There is evidently some confusion here, and the eglantine of one poet is not the eglantine of another. Sir Walter Scott, we take it, is thinking of the wild clematis or virgin's bower, when he wishes the eglantine to remain untrimmed. And Milton undoubtedly refers to the honeysuckle, which, twisting round the framework of a cottage-porch, tempts the neighbouring bees to rifle its calyxes of their honeyed sweets. But the *true* eglantine of our earlier poets seems to have been the prickly sweet-briar, formerly called *Rosa eglantina;* now known as *Rosa rubiginosa*. No plant is of greater value for a garden hedge, owing to the delicious fragrance exhaled not only by its flowers but by its leaves.

On the other hand, the "lush woodbine," which so often finds honourable mention in our poets, is none other than the honeysuckle, the "twisted eglantine" of Milton. Its botanical name is *Caprifolium*.

The eglantine and the woodbine, therefore, though occasionally confounded by careless writers, are two entirely distinct plants; the former being the sweet-briar of modern gardens, and the latter the honeysuckle.

It has been justly said by a writer (whom we have already quoted), that of all the flowers which, towards the end of summer and the beginning of autumn, adorn our pastoral scenery, "filling the air with fragrance, and the earth with beauty," none are more generally attractive than the wild climbing plants of our leafy hedgerows. By interlacing their delicate boughs, covered with foliage and flowers,—or with berries bright and sparkling,—or, as in the wild clematis, crowned with the lightest, feathery seeds,—they wind about the trees and bushes in festoons and wreaths of the utmost

Fig. 61.—"Our leafy hedgerows."

elegance,—and contribute in no slight degree to the aspect of richness and beauty which the landscape exhibits at this time of the year. As their stems are so slender and delicate that they would be crushed by the burthen of their flowery clusters and numerous leaves, or rent and uprooted by the wind, unless they found support from other plants, we see them hanging by their tendrils, or by their pliant arms, about the trunks of aged trees,—the ancestral elms, or "those green-robed senators of mighty woods, tall oaks,"—like a frail maiden to the sturdy arm of some strong-shouldered brother, or, it may be, of some one "nearer and dearer still."

In reference to those climbing plants, one curious circumstance deserves to be noted.

Some of them follow the sun's apparent course, that is, from east to west,—and always twine around the stem which supports them in the direction of left to right. Such is the case with the common black briony, so common in our woods and groves.

Others invariably twine *contrary* to the sun, or from right to left; as is the case with the convolvulus, or large white bindweed.

This singular tendency, be it observed, is always constant in each individual of the species, and if you endeavour to train one of these plants in a different direction, you will infallibly kill it.

The convolvulus will not grow from left to right, and the black briony will not grow from right to left. *Crede experto.*

The convolvulus, or white bindweed (*Convolvulus sepium*, or *Calystegia sepium*), is one of the most elegant, though one of the commonest climbing plants which festoon our willows, or creep over our grassy banks, or wind in and about our hedges. Its large white bells, which the country people unpoetically call "old men's nightcaps," are remarkable for their purity of hue and exquisite beauty of outline ; and the leaves, which are heart-shaped, equally claim our admiration. Like the pink field-convolvulus (*Convolvulus arvensis*), or the rosy-hued seaside bindweed (*Calystegia soldanella*), it is very tenacious of life, and if it once secures a footing, is eradicated with difficulty. Hence it is dearer to the poet and artist than to the farmer and gardener, each of whom pursues it with a determined hostility.

The Convolvulaceæ form a distinct family or order, containing forty-five genera, and upwards of seven hundred species. They are found in temperate and tropical countries; and include the dodder, sweet potato, scammony, spomœa, and the jalap plant.

METAMORPHOSIS.—A PHYSICO-PHILOSOPHICAL MEDITATION.*

If we are to understand by the term *metamorphosis* simply "a change," it is evident that everybody undergoes metamorphosis, is changed or transformed; nothing *is*, all *becomes*.

The water which flows on for ever, but never twice washes the same pebbly bed, will afford us an apt image of this perpetual "to become."

But even the said pebbly bed, like the hardest rock, like the seemingly everlasting granite, must and does change. The compact, chrystalline, azoic rock, without a trace of life in its dense mass, would eventually decompose if constantly assailed and affected by the moving waves of that gaseous ocean whose bed is formed by the terrestrial crust. If the rock is found covered by more or less stratified layers, its presence in the

* This section is translated from M. Hoefer, without addition or alteration

bosom of the earth will attest to passing generations the primordial incandescence of our planet at some epoch when life as yet was *not*,—when the liquid element, hurled far away into space under the form of vapour, exhibited the aspect of a "bearded meteor," or a comet, with blazing nucleus and incandescent tail. Many the changes which since that distant epoch have taken place upon the earth, and many more must occur before our planet ceases to contribute its strain to the grand harmony of the spheres. Our world will *end* as surely as it once had a beginning : its duration, though it be computed by hundreds of thousands of years, is nothing, will be nothing, compared with that of the revolution of yonder sun, circling, with its wondrous train of planets, around some mysterious centre as yet unknown. And in this period, hitherto incalculable, what chances of perturbation will necessarily arise?

Let us suppose that the centre around which oscillates, on the one part, the moon while drawing near and receding from the earth; on the other, the earth while drawing near and receding from the sun: let us suppose that these centres oscillate in the same manner around other centres as yet undetermined,—and this hypothesis is very rational, since it is based on the principle that everything moves or changes,—it may happen that in these periodical oscillations, one or more of the circulating masses will eventually fall into their focus of attraction, or will start so widely astray that the wheelwork of our world, the various parts of our planetary system, will separate,—not to be annihilated, for nothing in the universe can be *annihilated*,—but to be metamorphosed, and group themselves elsewhere in a different order.*

> * From these slow movements, which have been designated the "secular inequalities," we might with some probability infer the end of the world, which even Newton regarded as certain,—at least, unless "the Great Architect at times retouched His work." The inequalities or secular variations affect the elements of the orbits, such as the inclination of the plane of the orbit,—the semi-major axis of the ellipse, or the mean distance of the planet from the sun,—the eccentricity of the ellipse, or the relation between the distance which separates the forces from the centre and the semi-major axis assumed to be unity,—and the movements of the perihelions and the nodes. These elements change with extreme slowness. Thus, the inclination of Jupiter diminishes by 8″ in a century; and that of Saturn increases by 9″; but the very ecliptic varies,—it diminishes 33″ in a century.
>
> The variations of the eccentricities are scarcely computable by centuries; their effect is, that the ellipses insensibly approach or recede from a circular form.
>
> It is demonstrated by mathematical analysis that these variations are periodical, and confined within narrow limits, in such wise that "the planetary system can oscillate only round a certain mean, from which it never departs except by a very small quantity." But may not this very mean, which we have taken to be constant, oscillate round a more distant and still more constant mean?
>
> Observation had long ago detected a continual acceleration in the movement of Jupiter, and a not less certain diminution in the movement of Saturn. Now, to say of a star that its velocity augments, is to declare that it draws nearer its centre of movement. To say that its velocity decreases, is to affirm that it is retiring from that centre.
>
> It would seem, therefore, that Jupiter,[1] the greatest of our planets, is destined to be swallowed up, or absorbed, in the incandescent mass of the Sun, while Saturn,[2] with its belt and eight satellites, will gradually wander

---

[1] The equatorial diameter of Jupiter in English miles is 56,065; its density is ·24 of the Earth's; its distance from the sun, 494,270,000 miles; inclination of its orbit to the ecliptic, 1° 18′ 51″.

[2] The equatorial diameter of Saturn in English miles is 79,147; its density, ·12, the Earth = 1; its distance from the sun, 906,200,000 miles; and the inclination of its orbit to the ecliptic, 2° 29′ 36″.

It is thus that in chemistry, which I would call the *astronomy of atoms*, it is shown that bodies are only so far decomposed as to admit of their recombination in new forms; the end of one is the beginning of another.

Now, that which is true of the systems of the elementary bodies composing terrestrial matter, is, in all probability, true also—why should it not be?—of the systems of the celestial bodies.

Differences of magnitude, of space, and of time, which overwhelm our feeble imaginations, vanish before the unity of plan of the Creator's thought. A crystalline molecule, which will

further and further into the mysterious infinity of space. It is true enough that these catastrophes are very distant,— distant to a period of time which the human imagination cannot even grasp,—and the "common herd" will certainly feel no anxious apprehension about an event which will not take place until myriads of years have elapsed.

Yet, in the last century, the question excited the curiosity of certain scientific societies, and they directed the attention of the geometers to these formidable perturbations. Euler and Lagrange spent their keen intellects upon them to no profit. Laplace discovered that between the mean velocities of Jupiter and Saturn the ratios are simple, and capable of calculation; five times the velocity of Saturn perceptibly equals ten times the velocity of Jupiter. These terms, which, in the regularly-decreasing and indefinite series, might have been neglected, have acquired a value which was worthy of being taken into consideration. From thence result, in the movements of the two planets, those perturbations whose complete development necessitates a period of upwards of nine hundred years. . . . . This will be, then, another periodic inequality.

But, independently of the centres, which we suppose to be in themselves variable, may there not exist, in the space traversed by our planet, some cause of perturbation? Is our system so isolated in the universe that it neither receives nor loses aught of that which constitutes its force and matter? Is there no solidarity between its different worlds? And if this solidarity exists, is not their transformation a necessity?

not affect the finest balance, is a world, with an equator and poles of its own, and its central atom round which atomic satellites gravitate. Whether these atoms are infinitely small or infinitely great, whether the time of their revolutions is measured by thousandths of a second or by myriads of years, is of little importance so far as their gravitation (*ponderation*, or *poising*) is concerned. For this ponderation is absolutely identical, whether we call it *affinity*,—when speaking of the atomic movements of chemically decomposable matter; or *gravitation* or *attraction*, when referring to those atoms of the great whole which we call stars, and whose metamorphic scale is far beyond the range of beings planted on the surface of one of the stellar atoms. However profound may be the researches of our astronomers, they will never attain to a knowledge of the metamorphoses of worlds. The spectacle of celestial spheres rising anew from their ashes, like "the Arabian bird" of fable, will be as impossible for them as the knowledge of the decompositions and recompositions of our material bodies would be for chemists, planted on the surface of an atom of carbon. How, from such a stand-point, could they contemplate the manifold forms of matter, and embrace at a glance all its changes? . . . Well, we are relatively as powerless as these imaginary denizens of an atom of matter, rooted as we are to the crust of a planet,—a molecule suspended in the eternal ocean.

What shall we now say of the forms and movements of *living matter?*

In the first place, that they are infinitely more varied and more changeful than those of inanimate nature. Next, that the difference between their metamorphoses is very wide. The eye can follow the transformations of a rock exposed to the decomposing action of the agents which surround us on every side. This action is calculable, and the elements which it has dissociated may be determined and weighed. The effects of the force, called either affinity or attraction, which maintains these elements united, are not beyond the range of our observation; tables of affinity, and of atomic weights, have been constructed, which enable the chemist to dominate over matter, just as the astronomer embraces the stars, the atoms of the world, by the law of universal gravitation.

But no sooner is matter interpenetrated by that mysterious force which we call life, than our most potent means of investigation suddenly cease to be efficient. Undoubtedly, you may analyse the seed before you sow it, and thus may ascertain that it consists of *carbon, hydrogen, oxygen,* and *azote.* But with the same elements attempt to recompose your seed, using exactly the same proportions as those you discovered in it; and if you think that your synthesis has been successful, ensure that your grain, once confided to the earth, shall become a focus of divers movements, giving birth below to the ramifications of the root, terminated by the spongioles,—above, to the ramifications of the stem, garnished with leaves, flowers, and fruits; finally, ensure that this aggregate of organs, multiplying millions of times the weight and volume of the seed, shall always and exactly reproduce the same type or the same species.

If, with your apparatus,—if, with the means at the disposal of humanity,—you should succeed in achieving all these marvels of nature; then perhaps you might settle the great problem of what life is,—whether an independent force, or a simple modification of an universal force, of which heat, light, electricity, and magnetism, will be but different modes of manifestation.

And yet, even in such a case, you must not boast too loudly of your power; for you will find it necessary to turn and return the term you have arrived at, in every direction; nor will it furnish you with the relation or the cause of the formidable progression whose *alpha* and *omega*, whose beginning and end, escape us so absolutely.

Whence, in all its interminable metamorphoses, whence comes life? Whither goes it? Were our world to perish, its ruins would not cease, in their apparently disorderd movements, to obey the law of universal gravitation : they would so group themselves as to form other worlds, resembling the system of which they were anteriorly the framework. But this force in no wise tells you why, when, and how life will make its appearance on these spheroids of revolution, which, in their state of ponderation, are attracted in the direct ratio of their masses and in the inverse ratio of the square of their distances.

This is not all. Man justly plumes himself on having arrived, by a process of experiment, at the following irrefragable axiom : that " the matter which serves for the movements of life renews itself, while the mould or form remains." You may even affirm, without appearing too adventurous, that the innumerable whirling globules which enter into the composition of the human blood are so many microscopical individuals, each with its own proper life,—infinitesimal forms which are born, and move to and fro, and disappear, and are renewed, without the individual—whose aggregate they perform—having any consciousness of all this activity.

Thus it is that the collective integral being which we call humanity, lives and is developed through the removal of the individuals composing it,—ephemeral creatures, each of whom thinks himself a god!

Must we stop there? That would be to declare humanity the last term of a progression whose commencement and end, according to our own acknowledgment, completely escape us: it would be at once a contradiction and a flagrant violation of the great law of infinite continuity which reigns everywhere.

To suppose that beyond humanity there is only nothingness, would be to enunciate an hypothesis as puerile as that which pretended the earth was not only the centre of our system, but the sole inhabited or inhabitable point in the immensity of space, and that the stars of the firmament were created for the service and pleasure of mortals. Suppose that this absurd belief were true; of what use, I ask you, would be all our agitations, all our flutterings, all our conceptions, all our conquests, all our glories, all our memories, when the end of the world would sweep away and annihilate our race? It was well worth the trouble, truly, of being born, of living, and of suffering, to terminate, after all, in so inglorious a fashion! . . . Adhere to your hypothesis, materialist, if you have the courage; surely, no man of sense can accept it!

Let us now resume the thread of our meditations.

The end of our system has come at last: the sun, the planets, and their satellites form but one chaotic igneous mass, —a brilliant fugitive luminary, new-born to the inhabitants of worlds which have escaped intact.

The dust of our extinguished world will not be scattered hap-hazard; the molecules of matter, indissolubly linked together by universal gravitation, will so arrange themselves as to constitute a new, and perhaps a more perfect world. But in the constitution of this new world, balanced like the old, our human bones, our ashes united with those of our ancestors, may have, as far as they are matter, their due share. As for the Thought which makes the true power of humanity, which gives to man all his value,—Thought, perfectible and transmissible,—it will contribute nothing, because it is absolutely imponderable and impassible. Will it then be lost for ever?

If the world is to last for ever, you may justly regard as immortal the indefinite transmission of Thought, and the perpetuity of the memory of certain great men. But will all this avail, if the world must perish?

The world will never end, you say; it is eternal.

But how do you know this? If it has had a beginning, as geology and astronomy prove,* it will also have an end. This end, however, will not be an annihilation; it will simply be, as we have already pointed out, a transformation of matter. As for the problems, whether nature itself was created, and whether it is eternal, let us leave them to the discussion of heated theoricians, who are too blind to perceive that some questions it is wisest neither to affirm nor deny, but *to know how to ignore.*

The error, then, which we have been considering, destroys itself through its consequences. Let us admit, in effect, that our world—such as it is, just as it is—will last for ever. In that case what becomes of the power and travail of humanity? All they have accomplished are some slight changes of the terrestrial crust, barely sufficient, here and there, to modify the influences of climate. A limited number of men labour, it is true, for the progress and full development of transmissible thought. But even supposing that, in the course of centuries, humanity succeeds in comprehending, it can only *grow* through the development of the faculties of all its members, and the due balance of all the social forces by means of liberty. Supposing that reason, united to science and conscience, should finally combine in one family the various tribes and peoples scattered over the earth's broad surface; do you indulge yourself in the hope of crossing the limits of the human

---

* Our author, it must be understood, is considering these questions apart from the Scriptural standpoint.

organization, and establishing the royalty of man "through the interpretation and imitation of Nature?"

Do you cherish the idea of penetrating, through the perfect union of all your intellectual forces, the mysteries of creation?

No; you would never dare to form such a hope, to nourish such an idea, at least unless you felt that the earth (which you must first demonstrate) comprehends in itself the whole universe, that the humanity swarming on its surface is eternal, that every other creature is absolutely subordinate to it: in a word, that Man is all!

But we know how limited is human power. We are not masters even of the mechanism of the body; the movements of organic life are independent of our will; we can neither command the stomach, the lungs, nor the heart: that marvellous process of absorption and elimination, that perpetual movement hither or thither which constitutes the essence of the assimilative function, goes on in us—as in all living beings, animal or vegetable—completely outside our sphere of activity. Then, without quitting our planet, how numerous are the movements which still escape the human will!

It is quite different when we lift up our eyes to examine the face of heaven. We have no grasp whatever of the incommensurable spiraloids of innumerable worlds to which our own belongs; we have no means of communicating with the inhabitants of other planets; we cannot establish any interchange of thought with the *men* (if there be any) of Mercury, Venus,

Mars, Jupiter, Saturn,—who form, perhaps, like the *men* of the earth, the most elevated circle of material life, varying under an infinity of forms upon each of their floating domiciles. . . . I see you smiling, reader, because you do not believe that these *other earths*—satellites of the sun, like our own—are inhabited by beings analagous to our human race. You are at liberty not to believe it. But then, to be in agreement with yourselves, you ought to declare in favour of the Science of the Past, though demonstrated to be false, against the Science of the Present. Will you do so? Certainly not. But then, of two things, one: either you will be obliged to make the earth an unique exception, a kind of monstrosity in the midst of the other mechanism of the universe, which will be to throw yourself back upon the erroneous science of the ancients; or you must perforce admit that the earth is not specially privileged, and that the other planets, its companions, have also their human inhabitants.

Is this all? Alas, all this is nothing! The other worlds whose suns appear to us under the form of scintillating points or stars, will, undoubtedly, in like manner, possess their systems of planets and satellites. Why should they divaricate from the general plan of the universe? Now, multiply the number of the stars—who has counted them?—with the probable number of their planets, and you will gain, if this be permitted you, the number of *humanities* who people yonder star-sown space. And it is not only with *these* we must be able to correspond, but with the humanities of all the nebulæ of all the firmaments—for remember our starry heaven itself is but a nebula—that we must establish an interchange of ideas, if you would have your power, and civilisation, and intellectual royalty, something more than a mere optical illusion of your pride.

You do not cease to proclaim as an axiom that "there are no abrupt intervals in Nature;" that Nature never ventures upon sudden leaps or bounds ("in natura non datur saltus"), and yet you would make an exception for the world of thought —a world which no more lies outside the laws of nature than does the physical world. The former ought even to secure our preference; it is there only that we are free, that we can become true creators, by creating for ourselves our own happiness; that we can grow great before our own eyes, by following, not the tyrannous will of the *brute*, but the tender voice of the *angel;* by listening to conscience—that pure and infallible counsellor—conscience, the foundation of all justice, the latent force of generations passing away and coming, the universal gravitation of our species as of all the ultra-terrestrial humanities.

But how can the humanities, with which we suppose the universe to be peopled,—how can they communicate with one another?

The law of attraction comprehends not only the celestial bodies, but also their intervals, the intersidereal spaces. Do these spaces present a void to the thinking beings who probably people the stars?

Everything, force and matter, testifies to an entire unity of plan or thought, and the mind which is powerful enough to rise above all the attractions and influences of the body, the mind which, by its continuous labour, alone renders life and man of any importance, the mind once detached from the animal nature which it drags behind it like a prisoner to the chariot of his conqueror, shall it be inferior to inert matter? shall it be less than a ship without its compass? Continuous here, shall that continuity be elsewhere broken up? Surely this is impossible.

But how are we to recognise this continuity of essence in a spirit which, like man's, appears unavoidably fixed, like a parasite, to the surface of a planet?

Here lies the whole difficulty of the question; a question all the more perplexing because, in the search after scientific truth, the mind walks surely and steadily, except when resting upon the senses,—which are the backbone, so to speak, of the experimental method.

Answers, indeed, are not wanting, for each religion has its own. Every creed attempts to solve the problem. But then, faith is required to accept the answer or solution, and alas! faith is not implanted in every soul. It is useless, therefore, to wish that it might be the gift of those who, to the authority of tradition and long-established dogmas, prefer the liberty of discussion and the axioms of science. Are our bigots actually aware of what they do when they seek to compel into their circle of belief those minds which tend to escape from it at a tangent? We assert that in so doing they are guilty of an act of iniquity, of a veritable blasphemy.

Some explanation is necessary here. You believe, we hope, in the majesty, power, wisdom, and mercy of God, in the revelations He has vouchsafed to man, in the immortality of the soul. These are great problems, however; the greatest problems a mortal can venture to discuss. Already I see the bigot frowning; he professes to be shocked by the word "problem," he would fain substitute for it that of "certainty" or "truth." Well, through faith we accept them as truths; but, metaphysically speaking, they may be regarded as problems which the All-wise has submitted to man's earnest consideration. In fact, the God in whom you and we believe, in whom you and we put all our trust, has surrounded them with something of uncertainty, has invested them with so much of doubtfulness as may test our faith.

Yes, in doing so, He has had a purpose to fulfil. Let us think of a geometrician—and an ancient writer said that God, by creating the world, created geometry—for the sake of exercising the minds of his pupils, his children, giving them a problem to be solved.

If at the same time, he placed before them the solution, he would assuredly fail in his object. No means would remain of distinguishing the capable from the incapable, the studious from the indifferent, the idler from the worker, if they all found the question answered beforehand!

True, if the problem is too difficult, if the solution lies beyond the faculties of those whom he wishes to test and put to the proof, the master will not fail to furnish them with all the elements necessary for their guidance, whether they consider it from without, or whether they consider it with the help of their own inner consciousness.

But science and conscience stand in need of an equally difficult task; the first, that it may learn to observe clearly, the latter that it may learn to act purely. And it is here, above all, that the two-fold nature of man becomes a perplexity and a stumbling-block. On the one hand, man creates theories, in order to disembarrass himself of the science which calls for the exercise of powerful and laborious observation; on the other, he creates dogmas, which he hopes may lull to sleep that ever active, ever restless conscience, which demands fertile and beneficial actions, and rejects barren or deceitful phrases.

It is true that to many minds the discussion of the questions at which we have hinted seems a sorry work, because the time given for their discussion is necessarily so limited. What is life? they say. What can be effected in so short an interval? What can man hope to accomplish in the few short years that intervene between manhood, when the mind is mature, and old age, when the intellect grows enfeebled? These are the men who echo the old poet's mournful cry:—*

> " A good that never satisfies the mind,
> A beauty fading like the April flowers,
> A sweet with floods of gall that runs combined,
> A pleasure passing ere in thought made ours,

---

* William Drummond.

> An honour that more fickle is than wind,
> A glory at opinion's frown that lowers,
> A treasury which bankrupt Time devours,
> A knowledge than grave ignorance more blind,
> A vain delight our equals to command,
> A style of greatness in effect a dream,
> A swelling thought of holding sea and land,
> A servile lot, decked with a pompous name;
> Are the strange ends we toil for here below,
> Till wisest death make us our urns know."

But the poet, while taking this despondent view of life, forgets—not only that it is an opportunity, but—that it is the first stage of an eternal existence, and that the progress begun now shall be continued hereafter, when the mind, freed from its material clogs, shall enter upon the full fruition of its wondrous powers. And however brief it may be, is it not better it should be devoted to noble work than to ignoble idleness? Is it not better to use it as a time of preparation than to waste it in empty pleasures? To the despairing wail of the poet just quoted we would oppose, as far worthier of a gallant spirit, Ben Jonson's admirable conclusions:—

> " It is not growing like a tree
>   In bulk, doth make men better be;
> Or standing long an oak, three hundred year,
> To fall a log at last, dry, bald, and sere.
>   A lily of a day
>   Is fairer far in May,
>   Although it fall and die that night;
>   It was the plant and flower of light.
> In small proportions we just beauties see,
> And in short measure life may perfect be."

This is the true philosophy; to make our life as perfect as

our faculties will permit, and to look upon it as the introduction to a grander life, where the problems here discussed shall find a satisfactory solution.

> " Oh for the time when in our seraph wings
> We veil our brows before the Eternal Throne—
> The day when, drinking knowledge at its springs,
> We know as we are known."

Let us beware, however, when we devote our life to the pursuit of wisdom, that we do not make a false start. Do not let us diverge into the narrow way of bigotry and dogma. It is the property of exclusive and intolerant error to dominate and to reign alone. For this reason we systematically refuse and reject all control; and it is thus men have been fatally led to lay a rash hand upon intellectual liberty,—liberty, which the All-wise Himself has refrained from touching! Such is the early taint of our race, the moral situation of humanity.

But retribution has followed close upon this violation of all that is most sacred in our human world. The schools of philosophy, and the infallible creeds, founded upon authority to the exclusion of all mental and spiritual freedom, have never ceased to be at war with one another, and instead of labouring for that union which is strength,—that union so necessary to the happiness and advancement of humanity,—they have everywhere sown irreconcilable antipathies and bloody discords. Such is the religion of those who, under a pretence of worshipping God, worship only themselves! History affords us abundant illustrations of this melancholy truth.

Finally, let us return to our metamorphosis. The butterfly,

which the Greeks designated by the same word as the soul, ψυχή, springs, like every living creature, from an egg. But see what a transformation this egg undergoes! It becomes a caterpillar—a transitory form of animal life, remarkable for its voracity; this caterpillar is in its turn transformed; it grows into a chrysalis,—a temporary tomb, and whence issues the winged insect, alone adapted to the discharge of all the functions of a perfect animal. Gluttonous and greedy of enjoyment, the caterpillar lived for itself. So the caterpillar has no sex; while the butterfly hovers from flower to flower, has to seek therein its own nourishment, there to find the companion with whom its being is to be united.

This metamorphosis impresses the observer; principally because its periods are so distinct, and are so plainly marked by stages, which have all the appearance of veritable species. But he would greatly err if he thought it confined to a certain class of insects. All insects,—nay, more, all animals, including man himself,—undergo certain transformations in the course of their lives. Metamorphosis plays an important part in the unity of the general scheme of Creative Thought. If it is not always recognised, the reason is, that its phases are not boldly marked, that the periods blend into one another, that the various stages are effaced in the continuousness of the transformation.

But let not this continuity prevent the observer from detecting or discerning *in that which is, that which is to come.*

In the caterpillar he must learn to see the chrysalis; in the chrysalis he must be ready to recognise the future butterfly. And in all these changes the thoughtful mind may acknowledge a significant emblem of that immortality of the soul, that final transformation of humanity, which the Word of God has promised to us :—

> "Child of the sun! pursue thy rapturous flight,
> Mingling with her thou lov'st in fields of light;
> And, where the fields of Paradise unfold,
> Quaff fragrant nectars from their cups of gold.
> There shall thy wings, rich as an evening sky,
> Expand and shut with silent ecstasy!
> Yet wert thou once a worm, a thing that crept
> On the bare earth, then wrought a tomb and slept.
> And such is man; soon from his cell of clay
> To burst a seraph in the blaze of day." *

To assure ourselves by observation that, in living matter, there are organs irrevocably destined to decay or disappear, while others incline and grow towards perfection, is certainly one of the noblest studies imaginable. If philosophers, instead of employing their time in profitless speculations, devoted themselves to the examination of the great Book of Nature, God's second revelation, they would long ago have discovered what they are still seeking.

And we should now know how to distinguish, in man as in the insect, the rudimentary condition of his future life; and the belief in the immortality of the soul would not only be the creed of the Christian, but a scientific truth.

* Samuel Rogers.

APPENDIX.

APPENDIX.

THE SOLAR ECLIPSE OF 1870.

THE total Solar Eclipse, which is to render famous the month of December in the present year—famous, that is, in astronomical annals—deserves, we think, some special notice in our pages. In 1860, the *Himalaya* was fitted out by Government for the use of the *savants* desirous of observing the Solar Eclipse visible that year in Spain, and the results of the expedition were so important as fully to warrant the liberality of the Government. The eclipse of the present year will also be visible in Spain, but the path of the sun's shadow will pass as far north as Cape Spartel, and, crossing Algeria, will go northwards, *via* Sicily, and in the direction of Constantinople. The totality of the eclipse will not last so long as that of the Indian eclipse in 1868. The sun will be hidden from sight for no longer period than about two minutes twelve seconds, and whatever observations our astronomers are anxious to make must be made in that brief interval. It may not unreasonably be suggested that, in so short a time, no data can be ascertained worth the cost and trouble the proposed expedition will necessitate ; but the reader requires to be informed that the most valuable acquisitions lately won in the region of solar physics have been the result of observations which may almost be described as momentary.

What, then, is the important point on which our astronomers hope to gain information from a close examination of the approaching eclipse?

This question has been ably answered by a well-informed writer in the *Daily News*. The great problem to be solved is that of the strange appearance of the solar corona; of that glory of light which rings round the great luminary when totally eclipsed, to it, as some astronomers assert, a purely optical phenomenon, or, as others more reasonably declare, "one of the most imposing of all the features of the solar system." It is true that the former opinion is held by such men as Faye, Lockyer, and Professor Airy; but if it be founded on fact, the phenomenon loses all its importance, and nearly all its interest. If the other opinion prove correct, and the corona is discovered to be in reality an appendage of the orb of day, then the mind must at once acknowledge its unsurpassed splendour, its magnificent proportions. As its glow and ethereal radiancy often extends several degrees from the eclipsed sun, its diameter cannot be less than 2,800,000 miles.

Assume, then, that it is shaped like a globe, that it is the globular envelope (so to speak) of the sun, and we may conclude that its outer boundary would at most enclose a volume twenty-seven times as large as the sun's, or, in other words, *twenty-seven million times* larger than our terrestrial world.

It is in reference to this magnificent phenomenon that we hope to obtain some satisfactory information in December 1870. The results ascertained from the eclipse of 1868 proved to be almost directly contradictory to those obtained last year in the United States, and the problem now is, to discover *why* these results were contradictory.

In the meantime,—observes the writer already quoted,— some astronomers say that the observations already made suffice to show the real nature of the solar corona. "It has frequently happened that, with the means of solving a problem ready at their hands, astronomers have been content rather to wait till new observations removed their doubts, than to undertake the work of carefully analysing the facts already discovered. It is urged that the atmospheric explanation (Faye's and Lockyer's) is opposed by simple optical considerations; that the blackness of the moon's disc, in the very heart of the corona, affords the most unmistakable evidence that the moon lies nearer to us than the corona." The solar glare which, according to Faye's theory, illuminates our atmosphere, ought, according to this view—and ordinary reasoners will think this argument irrefragable—to cover the lunar disc as well as the surrounding heavens.

It is also argued, very forcibly, that the active atmosphere above the horizon of the observer is partially obscured during total eclipse, while all that part lying in the direction in which the corona is seen is wholly shielded from the direct rays of the sun, and cannot, therefore, furnish the "atmospheric glare" on which M. Faye relies.

M. Faye's theory is powerfully combated in a paper read by Mr Proctor before the Royal Astronomical Society, and a summary of which appeared in "Nature."*

He remarked that if we in reality possess sufficient evidence to determine whether the corona is or is not a solar appendage, it would be unfortunate for, and in some sense a discredit to science, if the precious seconds available for observers in December next were wasted upon observations directed to such a point determinable beforehand.

He proceeded to express his belief that the corona was no terrestrial phenomenon, arguing that the very blackness of the moon as compared with the corona (to which we have already referred), shows that the coronal splendour is *behind* the moon. In fact, the moon is projected on the corona as on a background; whereas, if the light be due to atmospheric glare, the corona ought to be a foreground.

This argument, however, may fail on the ground of its very simplicity. Mr Proctor, therefore, proceeded to inquire whether air which lies between the observer and the corona is really illuminated. He pointed out that all round the sun, for many degrees, perfect darkness should prevail if the illumination of the atmosphere by direct solar light were in question. As to the atmospheric glare caused by the coloured flames and prominences of the sun's disc, it must be comparatively small, bearing no higher proportion to the actual

"Nature," vol. i. p. 543.

light of the atmosphere than ordinary atmospheric glare bears to actual sunlight,—a very small proportion indeed.

But a fatal objection to the theory that the corona is due either to the glare from the prominences or to light reflected from the surrounding air, consisted in the fact that the so-called glare ought to cover the moon's disc.

Mr Proctor next referred to a number of observations in support of the view that the coronal light is not terrestrial; such as the appearance of glare during partial eclipses, the glare always trenching on the lunar disc; the relatively greater darkness of the central part of the lunar disc in annular eclipses; the visibility of that part of the lunar disc which lies beyond the sun in partial eclipses, the limb being seen dark on the background of the sky; and the visibility of the corona in partial eclipses;—were its most distinctive peculiarities, having been recognised when the sun's face is not wholly covered.

. What, then, *is* the actual nature of the corona?

May it not consist of the denser parts of meteoric systems travelling round the sun?

Leverrier has shown, and Mr Buxendell has helped to show, that the motion of Mercury's perihelion indicates the presence of a ring of bodies in the vicinity of the sun; but we have good reason for believing that for each meteor system our earth encounters, there must be millions on millions whose perihelia lie within the earth's orbit. Since the earth meets with fifty-six such systems, it will be seen how enormous probably is the total number.

Is, then, the corona simply the projection of these systems, illuminated as they must be by an intensity of solar light exceeding many hundred-fold that which illumines our summer days, nay, in certain places, were raised by his heat to a state of incandescence?

Such are some of the questions which, it is hoped, the eclipse of December 1870 will help to solve; and we think our readers will admit that their importance justifies us in looking forward to the approaching phenomenon with anxious curiosity. The bounds of human knowledge are gradually enlarging; and as they enlarge, they do but more vividly illustrate the infinite wisdom and perfectness of the scheme of creation.

COLOURED STARS.

Before bringing these pages to a close, and appending to our description of Everyday Objects (everyday, but not commonplace) the word "Finis," we wish to direct the reader's attention to the subject of coloured stars; not because these beautiful spheres can be fairly classed under our general title, but because they serve to show the peculiar attractiveness and novelty of scientific inquiry, and the wonderful results that spring from the persevering study of God's universe.

To the eye of the ordinary observer, one star seems much like another; these "lamps of night" differ, indeed, in brilliancy, in their distance from the earth, in their relative magnitudes; but to all appearance they exhibit a complete identity in physical constitution, and their white light has furnished poetry with one of its commonest images.

But the astronomer, armed with his wonder-revealing glass, has discovered that innumerable stars are exquisitely coloured; that some are blue, and others green, and others red. If on a clear night you examine Perseus with your telescope, you will find that the star Eta is of a glowing red. In the system of Ophiuchus you will encounter a blue star; in that of Draco, a deep red star; Taurus has a large red and a small bluish star; and in Argo, a blue sun is attended by a dark-red satellite.

The stars occur in double, triple, or multiple groups or clusters; and the systems so constituted differ from what we presumptuously call *the* Solar System in their colouring; and in this variety or diversity, yet another variety is visible. The binary coloured systems are not all composed of blue and red suns. In that of Gamma Andromedæ the great central orb is of an orange hue, the moon revolving round it green as emerald. What results from the marriage of these two colours—from this union of orange and emerald? Do we not see in it, says a French astronomer,[*] a combination, if the phrase is allowable, full of youth (*assortiment plein de jeunesse*)— a grand, a magnificent orange-coloured sun in the midst of the firmament—next to it, a resplendent emerald, which interweaves with the gold its greenish gleams?

The following lively picture is borrowed from the graphic pages of M. Flammarion.

[*] Flammarion, "Les Merveilles Celestes," pp. 160, 161.

There is more variety, he says, in the innumerable star-systems which crowd the fields of space, than in all the changes which the most skilful opticians can project on the screen of a magic lantern. Some of the planetary universes lighted up by two suns display the entire gamut of colours above blue, but are without the sparkling tints of gold and purple which contribute so greatly to the luminosity of our world.

In this category may be placed certain systems situated in the constellations of Andromedæ, the Serpent, Ophiuchus, and Coma Berenices.

Others are illuminated by red suns, as, for example, a double star in Leo.

Others, again, are wholly dominated by blue and yellow, or, at least, are kindled by a blue sun and a yellow sun, which afford but a limited series of the tints or shades comprised in the combinations of these primitive colours; such are the systems of Ceta, Eridanus (where one is straw-coloured, the other blue), Cameleopardalis, Orion, Rhinoceros, Gemini, Boötes (where the greater one is yellow, and the lesser a greenish blue), and Cygnus (where the smaller is "blue as a sapphire"). On the other hand, combinations of red and green may be found in Cassiopeia, Coma Berenices, and Hercules.

There are other stellar systems which, according to Flammarion, more nearly approximate to our own, in the sense that one of the suns illuminating them has, like ours, a white light,—the source of all colours,—while its neighbouring luminary casts a permanent reflection upon everything. Such, for instance, are the worlds revolving round the great sun of Alpha, in Aries; this great sun is white, but a smaller sun is constantly visible in the sky, whose azure reflections cover as with "a diaphanous veil" the various objects exposed to its rays. Number 26 in Ceta presents the same conditions, which are those, too, of a great number of the most brilliant stars. In the case of the worlds gravitating round the principal world of these binary systems, the originative white lustre would appear to create the infinite varieties noticeable upon the earth, with the distinction, of course, that an azure gleam constantly issues from the other sun; but for the planets gravitating round the latter, the predominant colouring will be blue, tempered by the action of the remoter white sun.

And just as white suns have blue suns for their companions, so have red suns yellow, or *vice versâ*. What a strange and curious radiance—like that of "a dome of many-coloured glass"—must be shed upon a planet by its red and green or yellow and blue suns! What striking contrasts, what fairy-like changes, must be produced by a red day and a green day succeeding alternately to a white day, or a day of darkness! Why do not our poets take up such a theme as this? Can they find aught more remarkable or more eëry in the dreams of their imagination?

Supposing that these wonderful spheres are surrounded, like Jupiter or Saturn, by a ring of satellites, what can be their aspect when lit up by several suns, each sun a source of different-coloured light?

This one, we may imagine, will be divided into hemispheres,

red and blue, that will suspend in the firmament a crescent of gold; another will seem to drop from the azure circle like a gorgeous cluster of emerald fruit. Ruby moons, opaline

FIG. 82.—"Bright with the beauty of the silver moon."

moons, moons of jasper and amethyst,—all shining with a mystic and indescribable radiance,—surely the heavens are decked with jewels like an Eastern queen!

Our poets may chant the praises of our modest, simple nights, of our calm, hushed heavens, bright only with the beauty of the silver moon, which pours its pale lustre with a winning charm on town and tree, on wood and lake. They may sing, as Barry Cornwall sings—

> "Now to thy silent presence, night!
>   Is this my first song offer'd: oh! to thee
> That lookest with thy thousand eyes of light—
>   To thee and thy starry nobility
> That float with a delicious murmuring—
>   Though unheard here—about thy forehead blue;
>   And as they ride along in order due,
> Circling the round globe in their wandering.
> To thee their ancient queen and mother sing. . . .
> Not dull and cold and dark art thou:
> Who that beholds thy clearer brow,
> Endiademed with gentlest streaks
>   Of fleecy-silver'd cloud, adorning
>   Thee, fair as when the young sun wakes. . . .
>   But must feel thy powers."

In some such ecstatic strains as this we may laud our moonlit nights, acknowledging in our heart of hearts the power of their silent, subtle loveliness; but how shall we compare them with nights made wonderful by a blending of golden and emerald fires? by the shifting coruscations of stars of many colours?

But here we must conclude a dissertation which threatens to become a rhapsody. It is difficult, however, to treat of such a theme, and to follow up all its strange and startling suggestions, in sober prose.

The Alpine Flora.

The Alpine Flora, of which in a preceding section we have given a very imperfect sketch, has been examined with loving minuteness by Messrs Elijah Walton and T. G. Bonney; and other united results of pen and pencil have been placed before the public in a handsome volume entitled "Flowers from the Upper Alps, with Glimpses of their Homes." A somewhat similar subject has been treated with much delicacy of feeling and fervour of description by the Rev. Hugh Macmillan, in his "Holidays in High Lands."

Among the Alpine plants sketched and described by Messrs Walton and Bonney are the beautiful *Lychnis*, a close relative of our corncockle and garden pink, and so called, says Gerard, the quaint old botanist, because it is a "light-giving flower." He also met with immense breadths of violet *pansies*,—"that's for thoughts," says Ophelia,—covering the sloping pasture-grounds of the Col d'Autune, and blending with rose and star gentians, soldanellas, primulas, and anemones. The pink blossoms and bare stem of the *house leek* is found "among the blocks tumbled down from the ice-streams, and its sparkling clusters seem to glow with a richer hue on the stony ruin which has shattered the pine, and crushed the life even out of the rhododendron."

In the Glarus Alps the odorous crimson tufts of the *Kammblume* are discovered in a profusion which rejoices and astonishes the traveller. The yellow-blossomed *ragwort* spreads its stem over the rugged mountain-sides in every Alpine district. But next in fame and beauty to the gentians, —which is, *par excellence*, the mountain-flower,—must be placed the *edelwein*, celebrated by Kobele and other famous poets, and growing in almost every part of the Alps from Dauphiné to the Dolomites.

For further details respecting a most interesting branch of botany, we refer the reader to the two works already named, and he will find that the barrenest, stoniest slope of the Alpine rocks, the bleak recesses where the ice river has its origin, and the rugged ledge where the pitiless winds seem to expend their wildest fury, have their objects of grace and beauty,— their gentle and ever-welcome evidence of the Divine love, the Divine wisdom, and the Divine power, as exercised for the delight of man in the remotest wildernesses of the earth. The love of God is everywhere.

INDEX.

"Academy," the, quoted, 369-371.
Achtysia, the genus, characteristics of 353.
Æshna grandis, described, 178; *forcipata*, 178, 179.
Agaricus Casareus, the, description of, 333, 334.
Agarics, varieties of, characterised, 335, 336.
Agarions, the, general characteristics of their form, 79.
Agarion virgo, the, described, 179.
Alps, the, an excursion among, 256-260.
Alpine snail, the, remarkable transparency of, 71.
Amanita muscaria, the, described, 328-330; its poisonous properties, 331-333.
Amanita solitarius, the, description of, 331.
Amphibia, the, number of species of, 361, 362.
Anaximander, the author of the "Uranometria," his enumeration of the stars, 25.
Ancients, the, their notion of the earth's shape, 104-107.
Anemones, the, blossoming time of, 157, 158; their sensitiveness to atmospheric changes, 158; varieties of, 159; references to, by Drayton, 159-161.
Animal life, distribution of, 357, 366.
Antipodes, the, existence of, believed in by the ancients, 107; proved by Columbus, 108, 109.

Arabs of the desert, the, their nomenclature for the stars, 22, 23.
Aratus, the author of "Phenomena," quoted, 22.
Arctic vole, the, discovery of, by M. Hugi, 50, 51; described by its discoverer, 51; examined and named by M. Martins, 51, 52.
Arcturus, the star, its colour, 23.
Argesilaus, the author of the "Uranometria," his enumeration of the stars, 25.
Aristotle, the Greek philosopher, his theory of the globe's sphericity, 106; on the mole's want of sight, 267.
Articulata, the, number of the species of, 362, 363.
Arum, the, described, 161, 162; its root, 163; its alimentary properties, 163; its various species, 164, 165.
Asterisms, origin of, 14.
Atmosphere, the, illumination of, 313.
Autumn, aspect of nature in, 366, 367.

Bailey, P. J., the poet, quoted, 3.
Balfour, Professor J. H., quoted, 208; on the diffusion of plants, 321-324.
Bearded pink. *See* Sweet William.
Birds, number of the species of, 359, 360.
Black Hellebore, the, described, 168.
Boehm, M. Joseph, his theory of the rise of sap, 136, 137.
Boletus, the, of the Romans, 334.
Boraginaceæ, the, characteristics of, 239-241.

Bosc, the French naturalist, his reference to the alimentary properties of the Arum, 163; his career sketched, 163.
Boyle, Robert, on the contemplation of the universe, 101, 102.
Bradley, the astronomer, his discovery of the nutation of the earth, 12.
Browning, Mrs E. B., her poetry characterised, 337; her "Aurora Leigh" quoted, 337, 338.
Bug, the harvest, description of, 349-352.
Burton, of the "Anatomy of Melancholy," quoted, 240.
Buttercup, the, its bulbous roots, 166; its stimulating properties, 167.
Byron, Lord, the poet, quoted, 101.

CÆSALPIN, the botanist, on the *Solanum nigrum*, 86.
Callisto, the legend of, 9.
Calypso, the island of, referred to, 19.
Calyx, the, of flowers, parts of, 211.
Campbell, Thomas, the poet, quoted, 236, 237.
Candolle, De, his opinions on the nature of red snow, 41, 42.
Cardinal points of the compass, how to ascertain our position in reference to the, 7, 8; process of determination of the, 17-19.
Carew, Thomas, the poet, quoted, 100.
Carnations, characterised by Jeremy Taylor and Drayton, 371; described, 372; varieties of, 373.
Carnivora, the, distribution of, 358.
Cassini, Jacques, his theory of the earth's form, 111, 112.
Castle pink, the, described, 375, 376.
Castelnau, the Count of, quoted, 363, 364.
Cereals, the, range of, 323, 324.
Cetacea, the, distribution of, 358; described, 208-215; curious forms of, 220-223.
Chaldeans, the, their conception of the earth's shape, 105.
Chamois, the, described, 56, 57.
Chaucer, the poet, his description of the daisy quoted, 138, 140.
Cheese mite, the, where found, 354; its ugly form, 355, 356.
Chomel, the botanist, on the medical properties of the daisy, 146, 147.

Chlorophyll, experiments in, 368, 369.
Chrysomela salicinia, the, described, 72.
Clare, John, the Northamptonshire poet, on the dragon-fly, 175.
Clematis, the, described, 168; celebrated by Keats, 168.
Columbus, Christopher, his demonstration of the existence of the Antipodes, 108, 109.
Compass, the, points of, 7.
Constellations, table of the number of stars in the northern, southern, and zodiacal, 27, 28.
Convallaria majalis. See LILY OF THE VALLEY.
Convolvulus, the, described, 383, 384.
Copernicus, the astronomer, his theory of the earth's rotundity, 110; antiquity of his so-called theory, 118, 119.
Cornwall, Barry, the poet, quoted, 229.
Corolla, the, of flowers, 214, 215, 223; varieties of, 225-229.
Crabbe, George, the poet, quotation from, 288.
Cynodon dactylon, the, described, 91-93.

DAISY, the, described by Chaucer, 138, 140; by Cowper, 139; by James Montgomery and William Browne, 140; its vegetation, 140, 141; described, 142-145; referred to by the ancients, 145; and by the botanists of the Middle Ages, 146; characterised by Wepfer, Tournefort, and Garidel, 147; its medical properties, 146, 147; George Withers' tribute to, 148; celebrated by Wordsworth, 149, 150; by Shakespeare and Milton, 151, 152.
Darwin, Erasmus, the poet, on the pimpernel, 261.
David's Chariot, position of, 8, 9.
Day and night, varying length of, 198, 199.
Delphos, the "navel of the world," 107.
Democritus, the philosopher, on the Milky Way, 25.
Dent de Jaman, the, described, 257.
Desor, M., the Swiss naturalist, his discovery of the snow-flea, 72, 73.
Desoria saltans, the, described, 73; *glacialis*, description of, 74.
Dial, floral, referred to and exemplified, 217, 218.

Diodorus, the historian, quoted, 93, 94.
Dioscorides, the author of "Materia Medica," his reference to Mercury's plant, 81; on the *Agrostis*, 95.
Diurnal movement of the stars, origin of its discovery, 16.
Dobell, Sydney, the poet, quoted, 129, 131.
Dogma, folly of, 292.
Dog Mercury, the, its power of propagation, 77, 78; described by Mr Sowerby, 78; how distinguished from weeds, 79; its diœcious character, 80; mentioned by Pliny the naturalist, 80, 81; and by Dioscorides, 81; its disappearance before the power of snow, 81.
Dog-star, the. *See* SIRIUS.
Dog's-tooth grass, the, characteristics of, features of, described, 91–93; its emollient properties, 96.
Draco, the constellation of, known to the ancients, 22.
Dragon-fly, the, described by the poets, 174, 175; its destructive properties, 175; the female, 182, 183; its metamorphosis, 183–187.
Drayton, Michael, the poet, his description of a spring-time posy, 159–161; on carnations, 371.
Drummond, William, the poet, quoted, 399, 400.
Dutrochet, the philosopher, his theory of the rise of sap, 136.

EAGLE, the, its affection for the mountains, 57; its flight described, 58–62; its nest, 62; the various species of, 62, 63.
Earth, the, shape of, as conceived by Homer and Hesiod, 103, 104; by Seneca, 104; by the Chaldeans, Anaximander, Anaximenes, and Zenophanes, 105; by Plato, 105, 106; by Eudoxes and Aristotle, 106; by the Greeks and Hebrews, 107; a problem, 398, 399.
Earwig, the, form of, described, 278–280.
Edelwein, the, in the Alps, description of, 418.

FLAMMARION, M., on coloured stars, 414, 415.
Flos Adonis, the, legend connected with, 169.

Forficulæ, the, group of, characteristics of, 286.
Forget-me-not, the, celebrated by Campbell the poet, 236, 237; described by Miss Pratt, 238; analysis of its form, 238, 239; legends connected with, 239, 240.
Friendship, the test of, 141.
Frigid zones, the, position of, 45, 46; the lizards of, described, 70, 71.

GALILEO, his discoveries in gravitation, 110, 111.
Ganges, the banks of, described, 248.
Garidel, the botanist, on the medical properties of the daisy, 147.
Garden-lily, the, described, 248.
Garden-nightshade, the, its extreme fertility, 82; description of, 83; by Tournefort, 84, 85, 89; its various forms, 85; its medical properties, 85–87.
Genesis, the book of, quoted, 191.
Gentiana lutea, the, its medical properties, 254.
Gentiana campestris, described, 256, 257.
Gentianaceæ, the family of, its characteristics described, 252–255; a Lilliputian specimen of, 255.
Gentians, the, of the Alps, description of, 10.
Gerard, the botanist, quoted, 418.
Gesner, Conrad, the botanist, on the tulip, 154.
Goethe, J. Wolfgang, the poet, quoted, 191.
Gold-crowned wren, the, description of, 64; its partiality to the society of other birds, 64, 65; its gymnastic accomplishments, 65; its cosmopolitan character, 65, 66; its smallness of size, 66.
Graetz's balls, origin of, stated, 55, 56.
Gramineæ, the, general characteristics of, 88–90; how described by the naturalists of the Middle Ages, 95, 96.
Gramen, the, Pliny the historian on, 94, 95.
Grasshopper, the, described by Leigh Hunt, 274.
Great Bear, the, origin of the story of, 9, 10; Homer's reference to, 10.
Grew, the botanist, his theory of the rise of sap, 124; on the calyx of flowers, 211.

HAY, aromatic, an instance of, 258.
Heat, the action of, described, 313-317.
Heavens, the, movement in, 115, 116.
Hebrews, the, their conception of the earth's supports in space, 107.
Hedgerows, the, wild climbing plants of, 382, 383.
Hepatica, the, described, 168, 169.
Herbs, the food of the primitive Egyptians, 93, 94.
Hipparchus, the astronomer, his discovery of the precession of the equinoxes, 116; his astronomical researches, 291, 292.
Homer, the poet, his reference to the Great Bear, 10, 19; and to Orion, 15; his scientific knowledge, 19; his conception of the earth's shape, 103, 104; quoted, 210.
Honeycomb, the, of the bee, described, 85.
Honeysuckle, the, celebrated by Scott, Burns, and Drayton, 380.
Hood, Thomas, the poet, quoted, 190.
Horace, the poet, quoted, 103.
House-leek, the, in the Alps, described, 418.
Houston, Professor, his table of the distribution of wheat and barley, 324, 325.
Howitt, Mary, the poetess, on the dragon-fly, 175.
Hugi, M., the Swiss naturalist, his discovery of the Arctic vole, 49-51.
Humanity, the future of, 393.
Human power, limitation of, 394.
Hunt, Leigh, his poem on the grasshopper, quoted, 274.
Huygens, the savant, his researches into the laws of gravitation, 111.

IMMORTALITY, an emblem of, in the metamorphosis of the caterpillar, 402, 403.
Infusoria, the, number of, 365.
Inglis, Henry, the traveller, quoted, 142.
Insects, the various orders of, enumerated, 280; the number of species of, 363, 364.

JONSON, Ben, the dramatist, quoted, 207, 248, 400.

KANE, Dr Elisha, the American explorer, his Arctic experiences, 39.

Keats, the poet, quoted, 29, 168, 190, 244, 298.
Kepler, the astronomer, his adoption of the theory of the earth's rotundity, 110.

LACTANTIUS, on the non-existence of the Antipodes, 108.
La Hire, the French botanist, his theory of the rise of sap, 134.
Landscape, a wintry, described, 5.
Leaf-wasp, the, its mode of depositing its larvæ, 71, 72.
Le Monnier, the French astronomer, on the theory of the nutation of the earth, 123.
Lesser celandine, the, celebrated by Wordsworth, 166; described, 166.
Libellula cancellata, described, 177.
Libellula grandis, described, 178.
Libellulæ, the family of, its characteristics, 174-183. *See* DRAGON-FLY.
Libellulites, the family of, its general characteristics, 179, 180.
Liberty, a plea for intellectual, 407.
Life, origin of, speculated upon, 390, 391.
Light, the nature of, 192, 193; physiological facts concerning, 194-196; the theory of, 197, 198; chemical action of, 312, 313.
Liliaceæ, the family of, its characteristics, 249, 250.
Lily of the field, the, of the Gospels, described, 242, 243.
Lily of the valley, the, described by Shakespeare, 242; its native countries, 242.
Little Bear, the, position of, 11, 12; distinct from the Great Bear, 12; first recognised by the Phœnicians, 20.
Little Vulcan, the, butterfly, described by Agassiz, 71.
Longfellow, the poet, his description of Orion, 13; quoted, 128, 129.
Lychnis, the, described, 418.

MACGILLIVRAY, the naturalist, quoted, 59, 60.
Macmillan, Rev. Hugh, quoted, 42-44.
Malpighi, the botanist, his theory of the rise of sap, 135.
Mammalia, the, summarised, 360.
Manilius, the poet, his references to the Great Bear, quoted, 20, 21.
Mant, Bishop, quoted, 63, 64.

INDEX. 425

Mariotte, the botanist, his theory of the rise of sap, 135.
Marmot, the, described, 53; its habits, 53, 54; the various species of, 55.
Mars, the planet, referred to, 301, 302.
Marsupialia, the, distribution of, 358.
Martins, M., the naturalist, his examination and description of the Arctic vole, 51, 52.
Marvell, Andrew, the poet, quoted, 217.
Matter, forms of living, 388, 389.
Mercurialis annua. *See* DOG MERCURY.
Metamorphosis, the function of, in nature, 401, 402.
Milton, his allusion to the daisy, quoted, 151, 152.
Mole, the, peculiar movements of, 265, 266; described, 267; Aristotle and Pliny on its want of sight, 267; its hands and fingers, 268; its favourite haunts, 269; marvellous properties ascribed to, 270, 271.
Mole-cricket, the, its form and habits described, 273-278.
Mollusca, the, number of species of, 362.
Moon, influence of, on the weather, 30, 31.
Moore, Thomas, the poet, his characterisation of the dragon-fly, 174; his allusion to the water-lily of the East, 246.
Mountain pink, the, described, 374, 375.
Mushroom, the, how to be distinguished, 327, 328; some varieties of, described, 329-337.
Mutability, the lesson of, 316.
Myosotis palustris. *See* FORGET-ME-NOT.

NATURE, beauty and suggestiveness of, 339; our imperfect knowledge of, 341, 342; the infinite diversity of, 365.
Nebria escheri described, 72; *chevrierii*, 72.
Newman Mr, author of "History of British Insects," quoted, 285.
Newton, Sir Isaac, his theories as to the form of the earth, 111, 112.
Nuphar lutea, the, described, 245.
Nutation of the earth, the, explained, 120, 121, 123; its discovery, 121, 122.

OBSERVER, the, of nature, 4, 5.
Œnothera biennis, described, 216, 218, 219.

Orion, the occultation of, its position in the heavens, 14; its place in the old mythology, 15.

PACHYDERMATA, the, distribution of, 359.
Parry, Captain, on the nature of red snow, 41.
Perianth, the, of flowers, 208.
Perrault, the botanist, his theory of the rise of sap, 135.
Petals, the, of flowers, described, 109, 224-227.
Phœnicians, the, their knowledge of the Great and Little Bears, 20, 21.
Physalis alkekengi, described, 222.
Pilgran, the meteorologist, his researches into the nature of climate, 31.
Pimpernel, the, all about, 260-264.
Pinks, the various kinds of, described, 374.
Planets, the, whether inhabited, 395.
Plants, discovery of the sex of, 207; appropriate to certain soils, 320, 321; diffusion of, 321-325; classification of, 343-345.
Plato, the philosopher, his notion of the form of the globe, 105, 106.
Pliny, the naturalist, on numbering the stars, 25; his reference to red snow, 39; and to the mercurialis, 80, 81; on the gramen, 94, 95; his reference to the daisy, 146; on the mole, 270, 271.
Plutarch, the historian, on the probability of the existence of the antipodes, 108.
Podura plumbea, the, described, 74, 75.
Polar Star, the, its position in the heavens, 18.
Poles, day at the, 200, 201.
Polygala vulgaris, the, described, 213.
Pointers, constellation of the, referred to, 22.
Pratt, Miss, on the forget-me-not, 238.
Proctor, R. A., on solar phenomena, 410, 411.
Protococcus nivalis, the plant described, 42-44.
Prunella, the, or "self-heal," described, 230-233.
Ptolemæus, his enumeration of the constellations, 24; his acquaintance with the so-called Copernican theory, 119; astronomical theories of, 297, 298.

Pythagoras, the philosopher, his familiarity with the so-called Copernican theory, 119.

QUADRUMANA, the, distribution of, 357, 358.

RAMOND, the naturalist, his researches in the natural history of red snow, 40, 41.
Ranunculaceæ, the, general characteristics of, 166-169.
Ray, the botanist, on the daisy, 146.
Red-billed crow, the, its appearance described, 68, 69; its habits, 69; the history of one which had been tamed, 69, 70.
Red snow, known to Pliny, the naturalist, 39; first described scientifically by De Saussure, 39, 40; discovered by Ramond in the Pyrenees, 40; by Captain Ross in Baffin's Bay, 41; described by Sir John Ross, 41; a fungus (?) 41.
Reptiles, the, distribution of, 361.
Rhizomes, or trailing roots, referred to, 88, 89.
Rhodius, Apollonius, the poet, quoted, 20.
Rodentia, the, distribution of, 358.
Rogers, the poet, quoted, 403.
Ruminantia, the, how distributed, 358.

SALAMANCA, the council of, referred to, 108.
Sap, the circulation of, 132; its ascent and descent, 133; theories concerning the rise of, 134-137.
Saussure, Benedict de, the naturalist, his observations on red snow, 39, 40.
Science, unselfishness of, 76; the peculiar characteristics of, 133.
Scutellariæ, the, characteristics of, indicated, 235, 236.
Seasons, the, changes of, 204, 205.
Seneca, the philosopher, his conception of the earth's support in space, 104.
Shakespeare, his allusions to the daisy, quoted, 150, 151; the lily described by, 242.
Shelley, the poet, quoted, 152.
Sirius, the star, its colour, 26; deleterious influence ascribed to, by the Greeks, 23, 24.

Snow, the, preservation of the germ of life in seeds and roots, 32; its composition, 32, 33; the reason of its preservative qualities, 34; in the form of crystals, 34; a reflector of light, 37; its utility to the agriculturist, 37; in the polar regions, 38, 39; red snow, *which see*.
Snow-bunting, the, description of, 66, 67; its favourite localities, 67, 68.
Snow, perpetual, the line of, where situated, 47; its variations, 48.
Snow-flea, the, its discovery narrated, 72, 73; its generic characters, 74.
Soapwort, the, described, 378; its juice, 378, 379.
Soil, effect of temperature upon, 317; cultivation of, 318-320; plants appropriate to different kinds of, 320, 321.
Solanum nigrum, the, described, 82-85.
Solar corona, the, nature of, 409-412.
Solar eclipse of 1870, the, considered, 407-412.
Solstices, the summer and winter, 45.
Somerville, Mrs, quoted, 289.
Sowerby, Mr, on the *Mercurialis annua*, 78.
Species, number of vegetable, 339, 340, 347, 348; number of animal, 356.
Spencer, Edmund, the poet, quoted, 2.
Spring, the awakening of, described, 127-131.
Staphylium olens, described, 271-273.
Stars, the, their infinite number, 4; their diurnal movement, 15-17; coloured, 412-417.
Stellaria, the, characteristics of, enumerated, 378.
Strickland, Miss, on the legend of the forget-me-not, 240.
Struve, Otto, his computation of the number of the stars, 25, 26.
Sun, the, movements of, in the heavens, 292-296; length of its radius, 296, 297.
Sweetbriar, the, celebrated by Spenser, Milton, Scott, and Drayton, 380; by Shakespeare, 381.

TAYLOR, Jeremy, on carnations, 371.
Telescopes, importance of their invention to science, 26.
Temperate zones, the, reference to, 45.
Temperature, effect of, on soil, 317.

INDEX. 427

Tennyson, Alfred, the poet, quoted, 6, 7, 57, 183, 247, 319.
Theophrastus, an ancient writer, on the *Agrostis*, 95.
Thomson, James, the poet, quoted, 32, 100.
Thought, freedom of, asserted, 311; indestructibility of, 392.
Toaldo, the Italian meteorologist, his researches into the phenomena of climate, 31.
Tournefort, the naturalist, his description of the *Solanum nigrum*, 84, 85, 87; quoted, 96, 234; on the medical properties of the daisy, 147.
Tragus, the botanist, his account of the *Solanum nigrum*, 86.
Triticum repens, the, described, 93.
Tschudi, M., the naturalist, his description of a red-billed crow, quoted, 69, 70.
Tulip, the, described, 152, 153; its introduction into Europe, 153; its cultivation, 154, 155.
Twilight, phenomena of, 205.

VERTEBRATA, the, orders of, 362.
Vole, the Arctic. *See* ARCTIC VOLE.
Voltaire, François Arouet, his popularisation of the Newtonian philosophy, 713.

WALTON, Isaak, his eulogium on the strawberry, 244.

Water, the crystallisation of, instanced and described, 36, 37.
Water-lily, the, described by Keats and Mrs Hemans, 244; by Wordsworth, 245; analysis of its form, 245; of the East, described, 246.
Water-ranunculus, the, description of, 167.
Winters, instances of some severe, 29, 30.
Winter-time, appropriate for the observation of celestial phenomena, 3; the passing away of, and merging into spring, 96, 97.
Withering, the botanist, on the *Solanum nigrum*, 86.
Withers, George, the poet, his tribute to the daisy, quoted, 148.
Wood-louse, the, its characteristics enumerated, 169-172; another species of, 173.
Wordsworth, William, the poet, his celebration of the daisy, quoted, 149, 150, his poem on the lesser celandine, 166 his reference to the lily of the valley, 245; quoted, 375.
Wren, the, described by Bishop Mant, 63, 64; its habits stated, 64.

YELLOW WATER-LILY, the, description of, 245.

ZODIAC, the, constellations of, 114, 115.
Zootoca pyrrhogastra, the, described, 70.

THE END.

www.ingramcontent.com/pod-product-compliance
Lightning Source LLC
Chambersburg PA
CBHW032140010526
44111CB00035B/631